CATASTROPHE
TO TRIUMPH

BRIDGES OF THE TACOMA NARROWS

RICHARD S. HOBBS

WSU PRESS

Washington State University Press
Pullman, Washington

Washington State University Press
PO Box 645910
Pullman, Washington 99164-5910
Phone: 800-354-7360
Fax: 509-335-8568
E-mail: wsupress@wsu.edu
Web site: wsupress.wsu.edu

Washington State University Press gratefully acknowledges the support and
assistance from the Washington State Department of Transportation in
making the publication of this book possible.

Library of Congress Cataloging-in-Publication Data

Hobbs, Richard S.
  Catastrophe to triumph : bridges of the Tacoma Narrows / Richard S. Hobbs.
     p. cm.
  Includes bibliographical references and index.
  ISBN 0-87422-289-3 (alk. paper)
  1. Bridges--Washington (State)--Tacoma Narrows--Design and construction--History.
2. Tacoma Narrows Bridge (Tacoma, Wash.) I. Title.

TG24.W3H63 2006
624.2'309797788--dc22   2006019947

Cover photo: *Narrow's Escape,* © Christopher Boswell
www.realwindow.com

Fine Quality Books from the Pacific Northwest

TABLE OF CONTENTS

FOREWORD

DOUGLAS B. MACDONALD
SECRETARY OF TRANSPORTATION

What is the power of an historical image that causes it to be permanently engraved in the popular imagination? Why does one image persist and become a recurrent cultural icon, while others slowly drift away, eventually becoming relics of the past?

Consider two events of failed engineering and technology that occurred in America more than six decades ago—one in 1937 and the other in 1940. Both were spectacular sensations, and reminders in their day that new technology can never entirely displace peril. Both images also are early testaments to the modern media's power to instantly send pictures of disaster around the world.

One, but not the other, involved a horrific loss of life, and thus might be thought to be the greater and more enduring image. That event, however, though hardly forgotten, has slowly receded into dimmer focus in human memory. It was the Hindenburg Disaster—on May 6, 1937, the world's greatest airship burned and crashed while approaching a mooring post at Lakehurst, New Jersey.

The other is the collapse of the Tacoma Narrows Bridge on November 7, 1940. It is so well-known today that just the span's nickname, "Galloping Gertie," conjures up images of a crazily swaying roadway and bouncing cables, the dismemberment of the deck, and the spectacular plunge of steel and concrete into Puget Sound, where the structural remains lie today. The loss of life was limited to the Coatsworth family's much mourned dog, Tubby. By miraculous good fortune, not a single human soul was lost. I often surmise that Galloping Gertie's story remains so popular because it was a disaster without deadly tragedy. Galloping Gertie probably tops the list of

America's best known—even best loved—civil engineering failures.

The great collapse has hardly lacked attention from engineers and authors. Henry Petroski, today's leading popularizer of progress in bridge engineering, has told the story of Galloping Gertie in more than one of his books. At the University of Washington, F. Burt Farquharson contributed technical and popular studies as a dedicated researcher of the collapse and its lessons. And there are many others, including fine treatments by Richard Scott, Martin Smith, Joe Gotchy, David Steinman, and Joseph Gies.

Thus, there has to be a good reason for a new book about Galloping Gertie. Why would the Washington State Department of Transportation, responsible for the design and construction of the most famous failed bridge in America, sponsor yet another publication on a so well worn topic? Is this evidence, perhaps, that there exists even in an institution a primal quest for exorcism?

A new book is welcome, even on this classic topic, as Richard Hobbs has gloriously demonstrated here. The opportunity has been seized to gather a wealth of never before published material, which has quietly waited discovery for fresh revelations to a fascinated audience. Furthermore, no previous publication presents even a fraction of the photographic documentation gathered here—an exciting springboard to a full consideration of both the 1940 and 1950 bridges and the quality of their architectural design.

Perhaps there is another simple reason (also central to the mission of this book) for the enduring hold of Galloping Gertie. The

Hindenburg disaster symbolized the end of air travel by dirigible, whereas Galloping Gertie was the beginning of the story of how the *next* suspension bridge would be constructed.

The building of a new span to replace Galloping Gertie taps another deep realm of curiosity. Who would draw the next design? Who would next venture to spin cable across the water? Who would erect the roadway out from the new steel towers to re-form the broken link? Would the new builders stand on the shoulders of their defeated predecessors? Would the bridge confirm lessons drawn from the bitter disappointment of failure?

Here is another tale worth telling, while at the same time, *Catastrophe to Triumph* provides a complete record of the striking shapes and forms of both structures—each a monument of mid-20th century engineering.

Then there is the *new* new bridge—the span being built today and scheduled for completion in 2007. This book's ominous title, *Catastrophe to Triumph,* needs comment, since this foreword is being written months before our newest Tacoma Narrows Bridge opens to the public. In reflecting on Galloping Gertie, surely we have learned this—do not tempt fate by trumpeting "triumph" before a great venture is truly finished. Triumph for the new bridge only will come when the span is completed and opened, and with proof it can stand against the gales and tides while safely serving the people crossing it for decades into the future.

We are pleased with our new book, and also hopeful and excited about our new bridge. We look forward with anticipation to the day when perhaps the newest bridge will stand as *another* triumph in spanning the Tacoma Narrows.

October 10, 2006

Acknowledgements

Richard S. Hobbs

On November 7, 1940, the central span of the Tacoma Narrows Bridge crashed into the wind-swept waters of Puget Sound. Within days, motion picture footage of "Galloping Gertie's" spectacular failure played to awe-struck audiences around the globe. Ever since, its story has fascinated people far and wide, passing into modern myth and legend, even as a grand new span arose at the Narrows in 1950.

Catastrophe to Triumph: Bridges of the Tacoma Narrows grew out of a public information obligation of the Washington State Department of Transportation (WSDOT) under the Environmental Impact Statement for the new 2007 Narrows Bridge and Section 106 of the National Historic Preservation Act to mitigate impacts to the integrity of the 1950 bridge. In 1999, WSDOT, United Infrastructure, the Federal Highway Administration, and the state Office of Archaeology and Historic Preservation signed a Memorandum of Agreement that laid the formal basis for development of historical information about the Tacoma Narrows bridges for the public at large.

Constructing this book, rather like a suspension bridge, required the efforts of numerous people. Their interest and passion for the history of the 1940, 1950, and 2007 Tacoma Narrows bridges provided me with inspiration and encouragement. Their contributions are acknowledged here with the deepest gratitude.

In the forefront are employees at the Washington State Department of Transportation. I especially thank Rick Singer, Business Manager for the Tacoma Narrows Bridge Project. Singer's unflagging belief in the value of *Catastrophe to Triumph* helped in a major way to bring it to reality.

Other WSDOT personnel who generously supported and contributed to this story include Linea Laird, Tacoma Narrows Bridge Project Manager; Vicki Steigner, Project Planning Engineer; Dawn Marie Moe of Interactive Communications; Tim Moore, Senior Structural Bridge Engineer; Cathy Downs, Librarian; Kip Wylie, Maintenance Supervisor at the Narrows; Jon Moergen, Maintenance Lead, also at the Narrows; Sandie Turner of Environmental Services; Filiz Satir, Public Outreach Manager; Sharan Linzy of the Bridge Preservation Office; John Johnson and Randy Powell of Engineering Records; Owen Freeman, Graphic Designer and Communications; and Claudia Cornish, Media Relations Manager. In addition to providing a wealth of facts and information, several of these people consented to personal interviews, adding another significant contribution to *Catastrophe to Triumph*.

Deserving an individual salute is WSDOT's Cultural Resources Specialist, Craig Holstine, whose meticulous cultivation of documentation on Washington State's historic bridges, including the Tacoma Narrows spans, substantially aided in preparing the final draft of this book.

Finally, I gratefully acknowledge the financial support from WSDOT and the leadership of the Secretary of Transportation, Douglas B. MacDonald, that ultimately made this book possible.

There are numerous other individuals whose knowledge and talents added to the successful completion of this story. My heartfelt gratitude first goes to Richard Scott, a gifted writer, part-time musician, full time planner for the Canadian government in Ottawa, and an internationally recognized suspension bridge authority. He read an early draft of the manuscript and offered a detailed and thoughtful critique. Scott's witty, insightful, and considerate suggestions were not only helpful, but inspirational.

A very special note of appreciation also is due to other people who kindly consented to interviews. All were closely connected to the history of the 1940 or 1950 Tacoma Narrows bridges. Earl White and Gerry Coatsworth Holcomb provided exceptional assistance well beyond my hopes and expectations, and they have my deepest gratitude. Others making invaluable contributions include Margie Brown, Howard Clifford, Arnie Colby, Jackson Durkee, Ed and Darcie Elliott, James Howland, Bill Matheny, Lewis B. Melson, Tim Moore, Charles "Chuck" Munson, Jean Robeson, Beverlee Storkman, Richard L. Swinney, and Jeanette Taylor. These names will come up again and again in *Catastrophe to Triumph*.

I also offer thanks to the following: Greg Anderson of Titus Will Enterprises; Christopher Boswell, photographer; Laurie Carron, architect; Michael Cegelis of the American Bridge Company; Sarah Clementson, artist; Chris Cunningham, computer consultant; C.W. Eldridge, son of engineer Clark Eldridge; Ed and Darcie Elliot of The Camera Shop; Eric DeLony, retired chief of the Historic American Engineering Record; Glenn Hartmann and Debbie Allen of Western Shore Heritage Services; James Howland, retired engineer; Dr. Tadaki Kawada, CEO of Kawada Industries, Inc.; Allan Larsen of COWI Engineers and Planners A/S; Doug McArthur; Robert Mester of Underwater Atmospheric Systems; Ray Plaut, the D.H. Pletta Professor of Engineering at Virginia Polytechnic Institute and State University; Steven Russell, author; Martin Smith, author and senior editor, *Los Angeles Times Magazine*; May Libby Smith; Dan Weber, structural steel inspector for AMEC Earth & Environmental; and Sharon Wood Wortman, Oregon bridge historian.

As a former archivist, I have an abiding appreciation for how curators, archivists, and other record-keepers assist researchers. I was fortunate to encounter talented, generous, and knowledgeable professionals at a number of archival repositories, historical societies, and libraries. Without exception, they were courteous and helpful in the best tradition of public service. First, I want to recognize the Washington State Archives staff, notably Dave Hastings, Patricia Hopkins, Mary Hammar, and Lupita Lopez. Equal gratitude goes to Bonnie Ludt at the California Institute of Technology; Vicki Blackwell of the Gig Harbor Peninsula Historical Society; Ernest White of the McChord Air Museum Foundation; Gene Morris of the National Archives and Records Administration; Bonnie McClosky at the Ohio County Public Library, Wheeling, Virginia; Marlene Wallin of the Oregon Historical Society; Pat Solomon and Robert Hadlow, Oregon State Department of Transportation; Lawrence A. Landis at Oregon State University Libraries; Tami J. Suzuki of the San Francisco Public Library, San Francisco History Center; Brian Kammens and Glenn Storbeck of the Northwest Room, Tacoma Public Library; Nicolette Bromberg, Hannah Palin, and Gary Lundell of the University of Washington, Special Collections; and Joy Werlink and Elaine Miller at the Washington State Historical Society.

I am indebted to my wife and soul mate, Lynette Dickson, whose steadfast support, patience, and loving acceptance of the many daily demands on my time made this book truly possible. She knows more about building bridges between people than I may ever know. I also wish to acknowledge, with a grateful heart, the unique contributions of my beloved children, Rhianna, Ryder, Ross, and Mariesa—my bridges to the future.

Finally, it is an honor to have the opportunity to formally say "thank you" to the exceptional staff at the WSU Press. The long process of shepherding a book from manuscript to print can be a rocky trail for a press and author. With expert editorial work and graphic design, and a good measure of patience and humor, the Press staff has been exceptional and most pleasant companions during this journey. I extend my hearty appreciation to Glen Lindeman, Kerry Darnall, Mary Read, Nancy Grunewald, Jean Taylor, and Caryn Lawton.

BRIDGING THE NARROWS

The bridging of the Tacoma Narrows is an uncommon story. Overshadowing the tale is the haunting specter of "Galloping Gertie's" collapse in 1940. The motion pictures and photographs of that spectacular disaster remain among the most compelling images in the public imagination.

Today, many of the technical explanations for Gertie's failure are widely known, especially to engineers and engineering students. So are the basic features of the replacement span, completed in 1950. Yet few people really know the full story behind Galloping Gertie's demise, or why the bridge was built in the first place. Even fewer know why it took a decade to erect a replacement, or the drama of the people involved in constructing both bridges. *Catastrophe to Triumph* tells those tales.

There is considerable published information about the 1940 and 1950 bridges. These sources, however, emphasize the 1940 collapse and often are partially focused, technical, and usually rely on a handful of sources. The only previously published work solely focusing on the Narrows bridges is Joe Gotchy's *Bridging the Narrows*, published by the Gig Harbor Peninsula Historical Society in 1990. Gothchy's book is a personal account by a worker who helped build both spans.

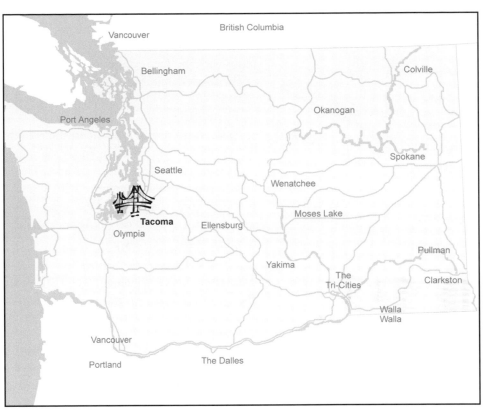

State of Washington, showing the Tacoma Narrows.
Owen Freeman, WSDOT

Indeed, the amazing bridge disaster provided some of the greatest lessons in how to build better bridges. Bridge historian Henry

NOTABLE MODERN SUSPENSION BRIDGES IN THE UNITED STATES

Bridge	Location	Length of center span (feet)	Year completed
Verrazano-Narrows	Upper and lower New York Bay	4,260	1964
Golden Gate	San Francisco Bay	4,200	1937
Mackinac	Mackinac Straits, Michigan	3,800	1957
George Washington	Hudson River, New York City	3,500	1931
1950 Tacoma Narrows	Puget Sound at Tacoma	2,800	1950
San Francisco-Oakland Bay	San Francisco Bay	2,310	1936
Bronx-Whitestone	East River, New York City	2,300	1939

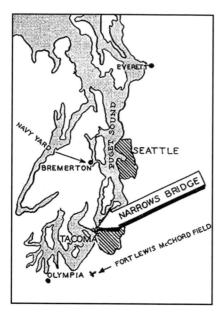

Prior to World War II, military planners considered the 1940 bridge to be an essential link connecting the Puget Sound Naval Shipyard with Fort Lewis and McChord Field. *WSDOT*

Petroski, in *To Engineer Is Human: The Role of Failure in Successful Design* (1992), has noted, "Failures, in fact, are a sure sign of progress."[1] Richard Scott's landmark volume, *In the Wake of Tacoma: Suspension Bridges and the Quest for Aerodynamic Stability* (2001), wonderfully tells a fuller story about the Narrows bridges. Scott's account of how engineers came to terms with the Tacoma catastrophe is thorough, detailed, and friendly to the average reader—a beacon amid the many publications on the subject.[2]

Studies also continue to appear today in the technical literature providing further insights into the mechanics of Gertie's collapse. These investigations provide engineers with a more complete understanding of these forces—lessons they can apply in designing safer, more efficient suspension bridges.[3]

THEMES IN *CATASTROPHE TO TRIUMPH*: *BRIDGES OF THE TACOMA NARROWS*

Catastrophe to Triumph presents a full account of the century-long effort to span and maintain bridges at the Narrows. In addition to a fresh view of the familiar events, much new information is added emphasizing the human aspect of this history. The text is formed around four basic themes—looking at the bridges as machines, art, and community connections, plus how people were (and continue to be) directly associated with the Tacoma Narrows spans.

Here you will meet designers and engineers, bridge workers with "nerves of steel," maintenance personnel, newspaper reporters, government officials, and commuters who relied on the bridges in their everyday lives. In today's world, with computers and other sophisticated equipment, it is hard to remember that bridge engineers in the 1930s and 1940s used little more than crude calculators, innumerable quantities of paper and pencils, and basic survey equipment to design and erect bridges.

The construction companies that built the marvelous 1940 and 1950 spans received considerable public notice, but not so the hundreds of craftsmen and laborers, whose sweat and toil erected the concrete and steel. Their names are virtually unrecorded, unless they fell victim during the construction process and were identified in newspaper accounts. And some did die, for the work often was dangerous. Their stories are included in *Catastrophe to Triumph*, as well as popular myths and mysteries that arose regarding the spectacular bridges standing in the stunning Narrows setting.

SUSPENSION BRIDGE BASICS

When viewed from a distance, suspension bridges almost appear fragile. Yet they are extremely strong and nowadays are the world's longest bridge type used for the farthest crossings. Typical lengths range from 2,000 to 7,000 feet. The longest suspension bridge in the world is the Akashi Kaikyo Bridge in Japan. Its main span between the towers measures an astounding 6,527 feet, and the bridge's total length is 12,828 feet.

Because they are relatively light and flexible, all suspension bridges are susceptible to the affects of wind. They vibrate and move, both vertically and laterally. The challenge for bridge engineers is to keep this motion within safe limits. Most vulnerable is the suspended structure, which supports the roadway. Engineers designed the 1940 Narrows Bridge to withstand winds up to 120 mph, but its center span collapsed in only a 42-mph storm.

THE PARTS AND HOW THEY WORK

A suspension bridge does just what the name implies. The "suspended structure" (i.e., the deck and stiffening truss or girder) is suspended from large cables that pass over the tops of high towers and are secured in anchorages at each shoreline.

The basic parts of a typical suspension bridge fall into two categories, the "superstructure" and "substructure." The superstructure consists of the suspended structure (deck, and truss or girder), towers, and the suspension cables. The substructure is composed of piers that support the towers, and the anchorages for the main cables at each end of a bridge.

The deck (road deck, or roadway) is where people drive and walk. It is continuous and supported by a truss, box girder, or plate girder. The suspended structure must be stiff enough to resist the forces of traffic loads and wind, yet be as light as possible. It is held by suspender cables hanging down from the main cables.

Large anchorages, or anchors, at both ends of a bridge act as counter weights holding the ends of the two main cables. Anchorages typically consist of concrete, though in rare circumstances solid rock is used. In an anchorage, the main cables splay into separate strands, which distributes the tension load evenly and safely in concrete.

The main cables stretch from either anchorage over the tops of the towers. The main cables consist of wrapped multiple strands, with each strand consisting of many parallel wires that are compacted. At the anchorage splay, each cable strand wraps around a strand shoe. Each of the strand shoes connects to an eye-bar, and the eye-bars are firmly cemented in the anchorage.

Towers support the main cables. The main cables pass over cable saddles at the tops of the

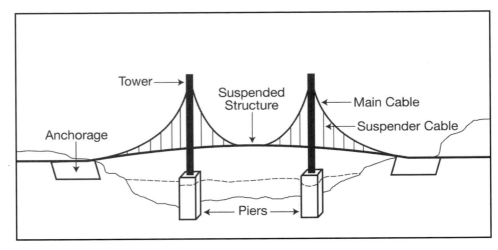

towers. At each cable saddle, the cable load is transferred from the cable to the tower. Towers are rigid enough to support the downward weight of the main cables and the upward force from the piers.

The suspended structure is attached to the main cables by smaller suspender cables, sometimes called "suspenders" or "hangers." Most of the bridge's weight (and any vehicles on the bridge) is suspended from the cables. The cables are held up only at the towers, which means that the towers support a tremendous weight (load).

Steel cables are strong but flexible, making long span suspension bridges susceptible to wind forces. These days, designers and engineers take special precautions to assure aerodynamic stability and minimize vibration and swaying in strong winds. The designed-flawed 1940 Tacoma Narrows Bridge proved to be the world's most famous example of aerodynamic instability in a suspension span.

Basic suspension bridge parts.
Owen Freeman, WSDOT

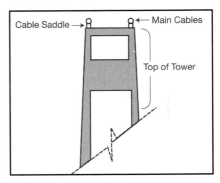

Notes

1. See also, Henry Petroski, *Engineers of Dreams: Great Bridge Builders and the Spanning of America* (New York: Alfred A. Knopf, 1995).
2. Another recent work that brings a broad perspective is Martin J. Smith and Patrick J. Kiger's *OOPS: Twenty Life Lessons from the Fiascos That Shaped America* (New York: Harper Collins, 2006). It includes the chapter, "The Preposterous Collapse of 'Galloping Gertie.'" The lesson learned, states Smith and Kiger, is "Ignore the past at your peril."
3. For example, engineering professor Ray Plaut and a team at Virginia Tech are developing new mathematical models to describe the initiation of torsional oscillations, the violent twisting that caused Galloping Gertie's failure.

Looking west toward the Kitsap Peninsula
across the 4,600-foot-wide Narrows, spanned
by the 1950 bridge. Visible in the mid-
distance are Wollochet Bay (center right) and
Fox Island (left), with the Olympic Mountains
standing on the horizon.
WSDOT

CHAPTER ONE

CROSSING THE NARROWS: IDEA AND DREAM, CA. 1888–1937

THE SETTING

In Puget Sound the archaeological record begins about 9,000 years ago. Native peoples typically located their villages beside rivers or along the shoreline on marine travel routes. Shellfish, salmon, and other aquatic resources were valued foods. They caught fish by gigging or netting, and in weirs and nets along streams and creeks. The pursuit of deer, elk, and bear also was significant. Plants, too, offered abundant variety in local diets.

Seaweed, plant fiber, ferns, acorns, roots, berries, and other vegetation provided medicines, food, and manufacturing materials for native households. Western red cedar was used in making plank houses, canoes, boxes, and baskets. The artifacts of daily life included a variety of hunting, plant, and fish processing tools, as well as bone, shell, and wood implements used in fabric and basket weaving.

The Tacoma Narrows was part of the traditional territory of the Puyallup people. The main Puyallup population was centered south of Vashon Island around Commencement Bay and the lower Puyallup River, including a tributary, Wapato Creek. The Puyallup also occupied villages nearer to the Narrows. One was called *swh'LOH-tseed*, located at the head of Wollochet Bay, three miles northwest of the southern end of the Narrows. Another, named *TWAH-well-kawh*, with six buildings in the mid-1800s, stood five miles to the north at Gig Harbor.

European explorers first came to the region in 1792. On May 20 of that year, a small group from the British expedition of Captain George Vancouver passed through the Narrows. One of them, a botanist named Archibald Menzies, scribbled his thoughts into a notebook, the earliest recorded words about the Narrows. Menzies wrote, "a most Rapid Tide from the northward hurried us so fast past the shore that we could scarce land." Vancouver named the southern part of the great complex of waterways that they were exploring after his lieutenant, Peter Puget, who led most of its mapping. Only later was the name "Puget Sound" applied to all of the vast channels and bays extending 90 miles south from Admiralty Inlet on Juan de Fuca Strait to Budd and Eld inlets near Olympia.[1]

The "Narrows" was named nearly a half century later in 1841, thanks to an American explorer. That year Captain Charles Wilkes led an expedition into Puget Sound. As he sailed southward past Point Defiance, Wilkes noted the strategic value of the bluffs overlooking the narrow channel, claiming that with gun batteries placed there he could "defy the navies of the world."[2]

By the mid-1850s, disease and conflict associated with Euro-American settlement had drastically affected native peoples. In 1854, U.S. government negotiations with the Puyallup, Nisqually, and Squaxin tribes resulted in the Treaty of Medicine Creek. The agreement forced a majority of natives to abandon their villages in southern Puget Sound and to relocate to reserves. But the restricted reservation lands dissatisfied many of them. In 1855–1856, the federal and territorial governments sent troops to contain the Puyallup and other tribes as part of a series of incidents that became known as the Puget Sound Indian War. Afterward, first in 1857, then in 1873, the U.S. government expanded the Puyallup Reservation near Tacoma to 18,032 acres.

In the years after these conflicts, white settlers came to the area in increasing numbers and settlements grew up around trade centers on the Sound. Tacoma was founded where the Puyallup River enters Commencement Bay. When the Northern Pacific Railway reached the area in 1873, Tacoma was destined to become a bustling commercial hub for lumbering, flour milling, and coal mining. Throughout the 1870s and into the 1880s, it grew into a thriving industrial center, resulting in incorporation as the City of Tacoma in 1884. East of downtown, the wide tide-flats became a grain elevator and warehouse district; other businesses sprang up too, including a salmon cannery and machine shops.

By the early 1900s, the "Narrows Strait," as it was then called, was a popular recreation area. In 1905, the federal government turned over Point Defiance to the City of Tacoma

for a park. Because houses and tents were prohibited in the park, people fished and camped along the Narrows from the southern boundary of Point Defiance Park to Titlow Beach. For the less well-to-do population in the area, the waterfront provided an inexpensive place to play, as well as to earn a little money from fishing. Year round, salmon, cod, squid, and octopus were caught in what was called a "fisherman's paradise" and sold in local markets. In 1910 much of the east shore of the Narrows was purchased by the Northern Pacific, which laid tracks along the waterfront. A small boat dock and resort, plus a general store, served locals.[3]

The turn of the 20th century saw Tacoma's leaders promoting the city as a major West Coast port, as well as the "Lumber capital of

An artist's birds-eye view of the Tacoma Narrows project site and its environs, 1938.
Washington State Archives

the U.S." Sailing ships and steamers busily cruised to and from city docks, daily passing through the Narrows on their journeys around the Sound. Today, Tacoma still is a leading center for the forest products industry in the Northwest.

West of the Sound lies the Olympic Peninsula, an 80-mile-wide wonderland rich in natural resources and abundant wildlife. Most prominent on the horizon are a range of great snow-capped mountains, topped by Mt. Olympus at 7,965 feet. Before the 1940s, the Peninsula had comparatively few residents, residing in small fishing ports, lumbering towns, and agricultural communities. Dense evergreen forests covered nearly three-fourths of the area, some two million acres. Beach resorts and mountain cabins attracted hunters, sport fishermen, and other tourists.

On the adjoining Kitsap Peninsula, the largest town was Bremerton, which was largely dependent on a U.S. Navy base located there. In the surrounding area, people earned a livelihood by fishing, small boat building, sawmill work, or in various agricultural enterprises, mainly in the production of dairy products, berries, vegetables, and poultry and eggs. Hood Canal had attracted a substantial number of people seeking summer waterfront homes. Gig Harbor was located about five miles north of the Narrows, a quaint fishing town in a scenic inlet. Boasting a postcard-like view of Mount Rainier, the town thrived on tourism.

Before there was a bridge at the Narrows, travelers to the Peninsula faced a long, slow, even costly journey. Access was either by roadways around the southern end of Puget Sound through Olympia, or by ferry from Point Defiance. By the late 1920s, with more and more people driving automobiles, demand grew for a better and more direct route. At the same time, many citizens came to consider the existing ferry system as "antiquated" and expensive.

The Narrows—at a little less than a mile wide—was a logical site for a bridge. But with a 200-foot depth and treacherous tides sweep-ing through at more than 8.5 miles per hour (12.5 feet per second) four times a day, it also was a challenging place to build. A bridge at the Narrows was long suggested, but construction difficulties and the high cost posed significant barriers.

BRIDGE IDEAS, CA. 1888–1927

First to suggest the notion of bridging the Narrows was John G. Shindler, a local rancher. When passing through the channel on a steamboat in 1888 or 1889, he said to the captain, while pointing to the bluffs on either side, "Captain, some day you will see a bridge over these Narrows." A half-century later when captain Ed Lorenz related the story of Shindler's comment to newspapers, he added, "We all thought Shindler was crazy."[4]

In 1889, the Northern Pacific briefly contemplated the construction of a bridge at the Narrows. A clerk in the NP's Land Department named George Eaton proposed a rail link (probably with a trestle rather than a traditional bridge) between Tacoma and Port Orchard, then the proposed site of the Puget Sound Naval Shipyard. However, there was little economic incentive to justify the effort. Ideas for a bridge then faded for more than 30 years.

The first real promoters of a Narrows bridge were neighborhood groups in Tacoma. In 1923, the Federated Improvement Clubs of Tacoma launched a campaign for a bridge, proposing to span the Narrows from Point Defiance. In late December, C.F. Mason, a realtor and president of the Federated Improvement Clubs, told newspaper reporters that his organization had been working on the project "for some months."[5]

The next proposal came three years later. In 1926, the Tacoma Chamber of Commerce endorsed a campaign for a bridge across the Narrows. Heading the effort was Llewellyn Evans, superintendent of the Tacoma City Utilities Department and president of the local Good Roads Association.

The first proposed bridge type for crossing the Narrows appeared in Tacoma newspapers in 1928. The sketch replicates the Carquinez Strait Bridge, a steel cantilever structure then under construction in California.

Also in 1926, Pierce County granted Mitchell Skansie a 10-year contract to operate an automobile ferry service across the Narrows. The contract guaranteed immunity from competition. By early 1927, Skansie's Washington Navigation Company began operating. The guarantee of "no competition" later became a significant impediment to building a bridge. In 1938, Thad Stevenson, the president of Tacoma's Chamber of Commerce, put it this way: "Every time we would get started with some financing, that contract would come up and we would have to include its purchase in the total, which was always too high as it was."[6]

BRIDGE PROPOSALS, 1927–1937

Between 1927 and 1937, civic groups in Tacoma and surrounding areas made numerous attempts to have a bridge built. In June 1927, the Roads Committee of the Tacoma Chamber of Commerce estimated that construction would cost between $3 and $10 million. Tacoma newspapers gladly trumpeted support and publicity, as momentum slowly began to build.

Meanwhile, proposals came from prominent engineers. Notable were Joseph B. Strauss from Chicago (who later built the Golden Gate Bridge) and H.H. Meyers of New York. Also, in 1928 a local civic activist and realtor, Charles A. Cook, suggested a 4,500-foot-long steel cantilever similar to the Carquinez Strait Bridge then being built near Berkeley, California. The $8 million price tag for the span seemed staggering at the time.[7]

In November 1928, the Tacoma Chamber of Commerce hired noted engineer David B. Steinman of the New York firm Robinson and Steinman to conduct preliminary work. Over the next two years, Steinman spent $5,000 on a survey, layout and designs, traffic estimates, engineering drawings, and reports. Tacoma's Sixth Avenue Commercial Club also rallied public interest.

In late December 1928, the Tacoma Chamber of Commerce gave its support to proposed legislation that would provide a state franchise for construction. City of Tacoma leaders and the Pierce County Board of Commissioners formally requested that Washington State officials erect a bridge. By early 1929, the state legislature passed a law authorizing a Tacoma Narrows Bridge and granted a franchise for the toll facility to three members of the Tacoma Chamber of Commerce—Llewellyn Evans, B.A. Lewis, and J.F. Hickey. The franchise, which ran counter to the licensed ferry service, was extended for two years in 1931 and again in 1933.

Meanwhile, David Steinman's design for a Narrows bridge appeared in local newspapers. Marvin Dement Boland, a Tacoma-based photographer, had taken a picture of the Narrows, on which a sketch of Steinman's proposed bridge was imposed. It looked strikingly similar to Steinman's vision for the "Liberty Bridge" that he was promoting at the same time in New York City (although neither would ever be built). Steinman's proposed suspension bridge would be 4,944-feet long with towers rising 670 feet above the Narrows. It featured a 2,400-foot center span, two side spans of 912 feet, and an approach span of 720 feet on the west side connecting to the Peninsula. The $9 million price tag seemed astronomically high.

Steinman's vision for a Narrows bridge remained in the imagination of area residents for nearly a decade, even as other proposals appeared in Tacoma newspapers as late as 1938. Today, the only known original copy of the photo—a framed depiction nearly five feet

long—can be seen at the Gig Harbor Peninsula Historical Museum. A museum volunteer donated the picture after receiving it from a friend, who had found it in the trash next to the Rosedale Community Club.[8]

Two years passed with little progress. The Tacoma Chamber of Commerce next decided that Steinman's firm was "not sufficiently active" and moved on to other ideas. Another proposal came in 1931 from Tacoma city engineers. They suggested a steel cantilever truss that would carry railroad traffic as well as motor vehicles and pedestrians. Designed by C.H. Votaw and Charles E. Putnam, the bridge would consist of five spans on four piers, with a 54-foot-wide deck allowing two lanes of highway traffic and a railroad track in the center. The $12 million estimate shocked almost everyone and the idea received more ridicule than support.[9]

Hope surged again in 1932, when the Tacoma Chamber of Commerce hired E.M. Chandler of Olympia. Chandler proposed a $3 million suspension structure with a 1,200-foot central span, a vertical clearance of at least 196 feet, and a deck 24-feet wide for two lanes of traffic. Chandler requested a loan to build the bridge from the newly created federal Reconstruction Finance Corporation (RFC). The cost would be repaid by tolls collected from users. One of the experts hired by the RFC to review Chandler's proposal was a noted New York consulting engineer, Leon Moisseiff, who would reappear in the Narrows bridge story in 1938.

The RFC refused to fund Chandler's idea. Too few cars and trucks would use the bridge, they said, and the cost to buy out Skansie's ferry system remained too high. Chandler announced a new plan two weeks later, made public on November 14, 1932. This time he proposed a 7,000-foot-long steel "cantilever," also with a 1,200-foot central span, and 6 other spans of about 600 feet each (plus approaches). The bridge would have 10 piers—2 of them on the shorelines, 2 out of water at low tide, and 6 piers in about 150 feet of water on either side of the center span. Chandler organized and incorporated the Tacoma Narrows Bridge Company with the cooperation

The photo taken by M.D. Boland with a superimposed sketch of David B. Steinman's proposed Narrows bridge, 1929. Tacoma is to the left, with the Gig Harbor side to the right.
Courtesy of Gig Harbor Peninsula Historical Society

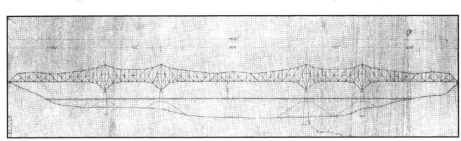

A steel cantilever proposed by Tacoma city engineers, 1931.
WSDOT

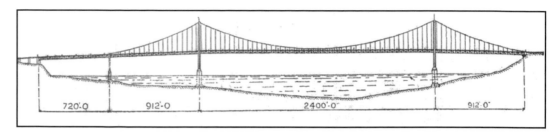

In March 1936, the New York engineering firm of Moran & Proctor prepared this preliminary design for a suspension span. *WSDOT*

of Pierce County officials and the Tacoma Chamber of Commerce. Evans, Lewis, and Hickey transferred their state franchise to build a toll bridge to Chandler's company.

The idea attracted little support at first. In February 1933, Chandler again applied for a $3 million loan from the Reconstruction Finance Corporation. By April, the RFC denied the loan request, again citing too little expected revenue from low traffic volume to make the project financially feasible. Chandler withdrew his application.

Pierce County governmental leaders now became encouraged by possible support from the Public Works Agency (PWA), which provided federal funds for major public construction projects. The PWA was one of the key parts of President Franklin Roosevelt's "New Deal" program to get the unemployed back to work during the Great Depression. In August

1933, Pierce County applied for a grant to build Chandler's steel cantilever at the Narrows, and, in late 1933, the Navy and Army declared their support for the proposal. The U.S. military clearly saw value in the project, because the bridge would link the Bremerton Naval Shipyard and Fort Lewis. A year passed with no action, however, before the PWA denied funding in October 1934.[10]

In January 1936, the War Department approved a revised application prepared by the Pierce County Commissioners. The new plan called for a suspension bridge costing an estimated $4 million; 45 percent of the money was to come from a PWA grant and 55 percent would be funded by Pierce County public utility bonds. By late March 1936, Tacoma newspapers announced that the New York engineering firm of Moran & Proctor had prepared preliminary plans for a 4,944-foot suspension bridge to cost $4,089,091.

Now, the "Narrows Bridge Gang" took up the cause. This coalition of supporters from the Sixth Avenue Business Men's Club, Gig Harbor Improvement Club, Young Men's Business Club, K Street Boosters, and other advocates in the area launched a statewide letter writing campaign to persuade President Roosevelt to support federal funding for a Narrows bridge. The group would meet weekly from January 1936 until the end of November 1938.

In the autumn of 1936, the area's congressional representatives also reenergized their support for the effort. For example, Senator Homer T. Bone, who had begun pressuring the federal government in 1933, strongly urged the PWA to support the bridge, calling it his "Project #1."[11]

The "Narrows Bridge Gang" in 1936. *Tacoma Public Library*

Driving Distances and Times Estimated for Vehicles Traveling at 45 mph, the Average Highway Speed in 1940

Route	Via Olympia	Via a Narrows bridge
Tacoma to Gig Harbor	107 miles; 2 hours 25 minutes	8 miles; 11 minutes
McChord Air Base to Bremerton Naval Shipyard	79 miles; 1 hour 45 minutes	39 miles; 55 minutes
Tacoma to Port Orchard	91 miles; 2 hours	35 miles; 47 minutes
Tacoma to Bremerton	90 miles; 2 hours	36 miles; 48 minutes
Tacoma to Port Angeles	152 miles; 3 hours 25 minutes	110 miles; 2 hours 27 minutes

BUILDING A CASE FOR A BRIDGE

In the 1920s and 1930s, the ever growing number of automobiles exerted a strong influence in the region, deeply affecting commerce and the American lifestyle and mobility. Citizens demanded new and better roads, with good bridges and fewer ferries. Furthermore, the Narrows project was being launched in a period when the ownership and control of toll bridges, ferries, and utili-

ties were shifting from the private sector to public ownership.

Tacoma's civic leaders saw the need to reduce the considerable driving distances between Tacoma and the Peninsula's communities as a compelling argument for a bridge. Peninsula residents felt equally enthusiastic about the concept—business and civic leaders promoted the benefits of better opportunities for trade with the more populated east side of the Sound.

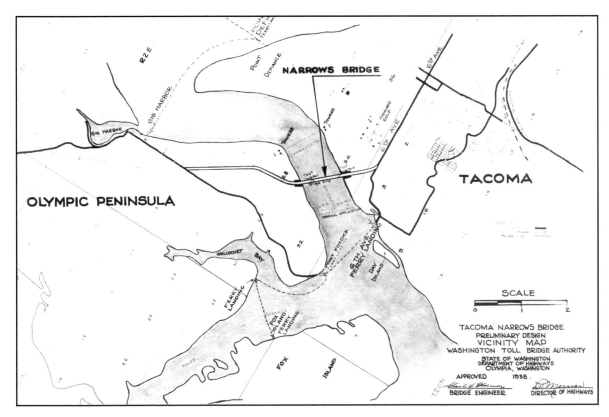

A vicinity map for the proposed Tacoma Narrows project, submitted with the Washington Toll Bridge Authority's application for federal funding, 1938.
WSDOT

Problems—Money and That Ferry

Up to this time, Washington had little experience in funding highway projects by selling bonds. The gas tax revenues that were being used to pay for road and bridge construction never accumulated enough funds for such large projects as a Narrows bridge. In 1933, when the legislature finally passed measures allowing bond sales to finance big projects, it set the stage for bridge proponents, but they still found it difficult to convince others that financing a Narrows span was a sound venture.

Advocates justified spending millions in bond funding by explaining that this cost would be repaid by user tolls. The critics, however, noted that the Peninsula had a small population, and the demand for travel to and from that area did not offer a strong incentive for a bond measure. Even when

Peninsula traffic significantly increased by the mid-1930s, federal officials still doubted the numbers presented to them by Tacoma and Peninsula promoters. After the onset of the Great Depression in the early 1930s, too, the financial hard times also proved to be a difficult hurdle for bridge enthusiasts.

The existing ferry service contract likewise remained a serious issue. Federal officials believed that Mitchell Skansie's Washington Navigation Company met local travel and commercial needs. Furthermore, the exclusive "no competition" franchise would not expire until 1936. Until then, any new plan had to include funds to buy out Skansie's ferries.

The end of 1936, however, would bring renewed hope to bridge promoters, as President Roosevelt's "New Deal" held promise for public funding of a Narrows bridge.

Notes

1. Archibald Menzies, *Menzies' Journal of Vancouver's Voyage, April to October 1792*, Archives of British Columbia, Memoir No. V, ed. by C.F. Newcombe (Victoria, B.C.: William H. Cullin, 1923), 33.
2. Charles Wilkes, *Narrative of the United States Exploring Expedition...*, Vol. IV (Philadelphia, 1850), 307.
3. Karla Stover, "The Narrows: Histories and Mysteries," *Tacoma Reporter*, April 29, 1999; "Trolling in the Narrows," *Tacoma Daily News*, August 1, 1905.
4. *Tacoma Times*, September 12, 1939.
5. "Federated Clubs Elect Officers," *Tacoma News Tribune*, December 8, 1923; "Start Big Drive Here for Bridge," *Tacoma Times*, December 29, 1923; T.A. Stevenson, "The Story of the Narrows Bridge and the People Who Made It Possible," unpublished memoir, Northwest Room, Tacoma Public Library.
6. "Fight for Narrows Toll Span Recalled," *Tacoma Ledger*, October 3, 1938.
7. "This Is How the Narrows Will Appear When Bridged," *Tacoma Ledger*, August 5, 1928.
8. "Action for Big Bridge Is Launched," *Tacoma Ledger*, November 22, 1928; "Tacoma's Proposed Narrows Bridge as Engineers Picture It," *Tacoma News Tribune*, March 5, 1929; "Start on Span Due Soon," *Tacoma News Tribune*, March 6, 1929; "Narrows Span Is City's Big Project, *Tacoma Ledger*, April 9, 1929. David Steinman was making regular visits to Portland, Oregon, at the time overseeing

construction of the St. John's Bridge over the Willamette River, which was completed in 1931.
9. "City Engineers Plan Narrows Bridge," *Tacoma News Tribune*, January 19, 1931.
10. Albert F. Gunns, "The First Tacoma Narrows Bridge: A Brief History of Galloping Gertie," *Pacific Northwest Quarterly* 72 (October 1982): 162–67; "Narrows Bridge Is in Sight," *Tacoma News Tribune*, November 15, 1932; "Starts on Narrows Bridge," *Tacoma News Tribune*, November 16, 1932; "Narrows Span One of Longest," *Tacoma Ledger*, November 30, 1932; "Puget Sound Bridge Proposed," *Engineering News-Record* 110 (February 2, 1933): 171; Articles of Incorporation, Tacoma Narrows Bridge Company, 1933, Washington State Archives; "Narrows Bridge Is Approved," *Tacoma News Tribune*, January 19, 1933; "Narrows Bridge Will Be Pushed," *Tacoma Ledger*, March 29, 1933; "Narrows Span Gets New Start," *Tacoma Ledger*, June 30, 1933; "Span Given Help in Olympia" *Tacoma News Tribune*, December 28, 1933 .
11. Gunns, 167; "New Hope for Bridge at Narrows," *Tacoma News Tribune*, January 22, 1936; "Narrows Span Now Tops List," *Tacoma Ledger*, March 29, 1936; "Narrows Span Plan Is Gaining," *Tacoma Ledger*, March 28, 1937; "Starts New Span Drive," *Tacoma News Tribune*, July 1, 1937; "Bone's Activity in Behalf of Bridge Told by Ickes," *Tacoma Times*, October 2, 1937.

CHAPTER TWO

CREATING THE FIRST NARROWS BRIDGE: 1937–1940

FACTORS FOR SUCCESS

From the outset, the key issue was whether the bridge could pay for itself. Private developers had believed that the traffic volume would be too low and there would not be enough toll revenue to pay for construction costs. Even the Public Works Administration and the Reconstruction Finance Corporation were skeptical. The economic downturn of the Great Depression limited financial options and stifled interest in new ventures. To almost everyone, the venture seemed "impractical."

Yet, by the late 1930s several factors came together that brought the required funding. In January 1937, the Washington State Legislature passed a law creating the Washington Toll Bridge Authority (WTBA). The legislature patterned the bill on a California law that led to the construction of the San Francisco-Oakland Bay Bridge and others. The governor, director of public services, state auditor, director of the State Highway Department, and the director of finance, business, and budget were named as members of the WTBA. The legislature also appropriated $25,000 to study the Tacoma-Pierce County request for a bridge. In June 1937, Pierce County transferred its building application to the WTBA. By this time, national and international events also were affecting plans for a Narrows bridge.

MILITARY AND STRATEGIC NECESSITY

A bridge at the Narrows was seen as a military necessity to link McChord Air Field south of Tacoma with the Puget Sound Naval Shipyard in Bremerton. In fact, successful funding would be closely linked with the nation's

defense strategy in the late 1930s. In particular, McChord Air Base became a catalyst in the effort to erect a Narrows span.[1]

McChord began as a municipal airfield in the late 1920s when the City of Tacoma purchased 900 acres for an airport. "Tacoma Field" opened in early 1929, and by 1934 it was one of the largest in the western United States. The U.S. Army soon became interested in Tacoma Field as a defensive base in the Pacific Northwest. At that time, Congress authorized the establishment of military airfields in six strategic areas of the country, and the Tacoma airport was selected for the Northwest region.

Proponents for a Narrows bridge now had strong support from the Army, because of its installations at both McChord Field and Fort Lewis, and from the Navy with its shipyard in Bremerton. In May 1938, Washington Governor Clarence D. Martin officially authorized the transfer of Tacoma Field to the federal government, and it became McChord Field.

The Board of Consulting Engineers for the the Washington Toll Bridge Authority, 1938. Left to right: Reginald Thomson, Luther Gregory, Ray Murray, Charles Andrew, and Ray McMinn.
Washington State Archives

The WTBA included a depiction of the Tacoma Narrows Bridge in its logo, 1938.
Washington State Archives

That summer, construction of the $5 million airbase began on a large scale. By 1939, the Air Corps facility included a 1,285-man barracks, radio transmission building, heating plant, hospital, two runways, four airplane hangars, and warehouses, plus facilities for maintenance, heating, water, electricity, and fire protection. In late June 1940, the first B-18, B-18A, and B-23 bombers would begin arriving.

With Italy invading Ethiopia in 1935, a civil war erupting between fascists and socialists in Spain in 1936, and Japan attacking China in 1937, the possibility of America's involvement in a world war played a role in the acquiring of funding for a Narrows bridge. After Adolf Hitler's rise to power in 1933, other war clouds began gathering over Europe as the Nazi government launched a massive military build up. In February 1938, Hitler appointed himself as commander of the German army. A month later, the Nazis annexed Austria, and, within months occupied Czechoslovakia as well. Meanwhile, the Roosevelt administration quietly bolstered the nation's military by funding major rearmament programs. At the same time, the administration poured millions of dollars into public works projects, including roads, dams, and bridges.

On May 23, 1938, two weeks after City of Tacoma officials transferred Tacoma Field to the federal government, the WTBA submitted an amended application to the PWA and applied to the RFC for a loan to build a suspension bridge across the Narrows.[2]

MONEY COMES—WITH STRINGS ATTACHED

The WTBA's application for federal support was prepared under the direction of the State Highway Department, headed by Lacey V. Murrow, who flew to Washington, D.C., to personally deliver the package. The proposal included a preliminary design by a brilliant and energetic state bridge engineer in his early forties, Clark Eldridge. At the time, Eldridge

also was designing the Lake Washington Floating Bridge, the world's first concrete pontoon span.

A Narrows bridge had been special to Eldridge since June 1937. Soon after Murrow had instructed Eldridge to prepare design and construction plans for the Lake Washington Floating Bridge, Eldridge asked about the Narrows site. "No funds," Murrow replied. Eldridge then queried about the $25,000 that the legislature had given to Pierce County. Murrow told him to try and get it. That afternoon, Eldridge secured the financing from the Pierce County Commissioners. At that point, the Narrows project became for Eldridge, as he later said, "My bridge." That was true on a personal level, but also on an operational basis. Murrow turned over the entire Narrows plan to Eldridge.[3]

Both the Narrows and Lake Washington projects were seeking funds from the PWA and RFC. Eldridge's Narrows' plan was what he later termed "a tried and true conventional bridge design." The state estimated the cost at $11 million.

Although federal authorities had been skeptical for several years about whether tolls could generate enough revenue to repay a loan, ultimately this did not block the project. The PWA agreed to a grant of 45 percent of the construction cost. However, the money was substantially less than the State Highway Department had requested. It also came with strings attached. The PWA required that the Washington Toll Bridge Authority hire outside consultants for the bridge design—but these consultants would be selected by the PWA.

Two years later, Highway Director Lacey Murrow made an official (but not public) reference to this troubling turn of events when he submitted an amended application for an increased grant to the PWA. In March 1940, Murrow listed the reasons why the price had risen on the project, including the following: "Engineering cost was greatly increased by the requirement of the Public Works

Administration that the owner engage the services of consulting engineers, namely Moran, Proctor, Freeman and Mueser [generally known as Moran & Proctor] and Leon Moisseiff, at a stipulated consulting engineering fee schedule aggregating 2½% of the contract cost." This fact was publicly echoed later (after the bridge's collapse) by Clark Eldridge, who told newspaper reporters, "We were told we couldn't have the necessary money without using plans furnished by an eastern firm of engineers, chosen by the money lenders."[4]

It remains unclear why the PWA insisted on such a requirement. No similar bureaucratic meddling complicated funding for the Lake Washington Floating Bridge, the WTBA's other Eldridge-designed project. According to Eldridge, "eastern consulting engineers" went to the PWA and RFC and said that the bridge could be built for much less than the $11 million Eldridge's design would cost. By "eastern consulting engineers," Eldridge meant Leon Moisseiff. According to Eldridge, Moisseiff had told the PWA that his changes would cut the estimated cost to less than $7 million. If true, Moisseiff's actions could be viewed as unscrupulous; regardless, his message that he could design a cheaper bridge fell on willing ears at the PWA.

At the time, the circle of consulting suspension bridge engineers was a small, competitive one. Since 1932 when he first reviewed E.M. Chandler's plans for a span at the Narrows, Moisseiff had an interest in the project. Furthermore, Moran & Proctor had been engaged by Pierce County as recently as 1936 to investigate suspension bridge construction at the Narrows. When Murrow appeared on the PWA's doorstep with plans drawn up by one of his own engineers, the implication was clear—there would be no big commissions for private consulting engineers. However, the PWA instead ensured that a consulting engineers' fee of 2½ percent (an estimated $160,00 to $175,000) of the contract cost would be split between Moisseiff's firm and Moran & Proctor.

Leon Moisseiff was one of the most eminent suspension bridge designers in the nation. He was renowned for his technical expertise and personal character; contemporaries considered him a person of high moral integrity. Considering his reputation, it seems possible that the "back door politics" that dropped the profitable redesign job at Moisseiff's doorstep was initiated not by him, but by PWA bureaucrats who wanted to save the agency some money.

On June 23, 1938, the PWA granted partial funding for the project (the PWA also helped fund the Lake Washington Floating Bridge). Soon, the RFC announced a loan to the state for the remaining funds. This marked the culmination of more than 14 years of community efforts.

The WTBA reluctantly agreed to the terms, and on June 27, 1938, accepted the PWA grant of $2.7 million and an RFC loan of $3.3 million. Leon Moisseiff was hired to design the superstructure, and the firm of Moran & Proctor was appointed to design the substructure.[5]

The WTBA faced a deadline of December 1, 1938, to submit detailed plans to the PWA. Events began to move rapidly. Lacey Murrow took a train to New York, bringing a contract to Moisseiff, who signed it on July 7, 1938. The next day, Murrow telegrammed Eldridge with instructions to send the designs, specifications, soil test data, and contract forms to Moisseiff and to Moran & Proctor. On July 18, just 10 days later, Moisseiff submitted his report to Murrow with his recommendations for changes to the superstructure. When Moisseiff's design arrived at the State Highway Department in Olympia, the agency's engineers protested, calling Moisseiff's plan "fundamentally unsound." The design would make the bridge proportionally lighter and narrower than any ever built, they said, "in the interests of economy and cheapness."[6]

At the time, none of this became public; the WTBA wanted the structure and the project went forward. Murrow wrote to engineer

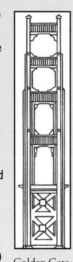

Headquarters for Clark Eldridge and his staff, about a mile from the bridge, 1939.
Washington State Archives

Charles Andrew in California, asking for copies of the plans for the San Francisco-Oakland Bay Bridge. The capable, confident, and soft-spoken Andrew had been the Washington State Highway Department's first bridge engineer (1921–1927) before moving to San Francisco. He was the principal engineer for the San Francisco-Oakland Bay Bridge, leading the design and construction of what was, at the time of completion in 1936, the world's largest suspension bridge. Now, Murrow asked Andrew to come back to Washington. Andrew agreed and assumed the position of chief consulting engineer for the WTBA.

In late July and early August, the involved parties met to inspect the site and discuss the project. Leon Moisseiff and his associate, Frederick Lienhard, both came, as well as a representative from Moran & Proctor. So did a consulting engineer hired by the RFC named Theodore Condron. A month after receiving Moisseiff's plan for the superstructure, the WTBA's Board of Consulting Engineers, headed by Andrew, issued their report. The August 31, 1938, document first reviewed Moran & Proctor's design. They found it "impractical and objectionable," and recommended substituting Clark Eldridge's original design for the piers, which the PWA approved. The plan submitted by Moisseiff was approved for the superstructure. Interestingly, in an anticipation of contractors' bids that might exceed available funds, Moisseiff telegrammed Murrow on September 24, suggesting that the 4,000-foot suspended structure be made of aluminum instead of steel, saving about $400,000. Eldridge declined, believing the saving was not sufficient to affect prospects for securing federal funds.[7]

On September 27, 1938, the state opened construction bids. The Pacific Bridge Company submitted the low bid. The Bethlehem Steel Company would be the associate contractor for supplying steel, while John A. Roebling's Sons Company of New York would supply the wire. All of the bids were above the budgeted projections, so the state reap-

plied for the extra funds needed. The PWA increased its grant to $2,880,000 while the RFC raised its loan to $3,520,000, for a total of $6.4 million. The design consultants, Moisseiff's firm and Moran & Proctor, divided the standard engineering fee of 2½ percent of the construction cost, amounting to $160,000.

Groundbreaking for the start of construction, November 23, 1938.
Washington State Archives

NOVEMBER 23, 1938, TO JULY 1, 1940

Construction began on November 23, 1938, although the official start date according to the contract was two days later. Clark Eldridge, heading the state's team of engineers, soon opened an office a mile away.

As work proceeded a growing throng of onlookers watched from the bluffs on both sides. From Tacoma, the Sixth Avenue route to the Narrows frequently became crowded. On Sundays especially, many area residents packed a picnic basket and found a spot overlooking the construction site where they could watch the progress.

A year later, as the piers, towers, and anchors were being completed, cable spinning was scheduled to begin on January 10, 1940. This novel feature of suspension bridge construction particularly attracted many sightseers. By the first week of May 1940, workers also completed the steel floor system, and by June finished pouring concrete for the roadway. Opening day was rapidly approaching.

A young woman poses on a cable saddle in a publicity photo, ca. August 1939.
Washington State Archives

RFC consulting engineer Theodore Condron (left) on a 1939 visit to the construction site with Clark Eldridge, David Glenn, Charles Andrew, and Mr. Reese.
WSDOT

Sunday visitors at the east anchorage, February 11, 1940.
Washington State Archives

CBS correspondent Edward R. Murrow (second from left) celebrates the opening of the Tacoma Narrows Bridge with workers, July 1, 1940.
Gig Harbor Peninsula Historical Society, Bashford 2709

"A dream come true," it was said on July 1, 1940. The day brought clear blue skies and a crowd of thousands to the opening festivities and official dedication. A two-hour parade in downtown Tacoma initiated the multi-day celebration marking both the opening of the Tacoma Narrows Bridge and McChord Air Field. Peninsula towns, from Port Orchard to Bremerton, sent floats and marching bands, and some 7,000 people lined the sidewalks to watch. Local newspapers called it the "biggest crowd that ever piled up in Tacoma Streets."

At the Narrows, a jubilant crowd surrounded the speaker's podium where politicians, citizens, and government dignitaries hailed the new span. They praised the bridge's beauty, proclaimed the military and economic benefits sure to follow, and extolled the natural wonders of the south Sound region. On hand were Governor Clarence D. Martin, some of the state's delegation to the U.S. Congress, Washington Toll Bridge Authority members, Lacey V. Murrow, and Colonel E.W. Clark, the acting commissioner of the Public Works

Opening day, July 1, 1940, was celebrated with speeches, military parades, and marching bands.
Washington State Archives

Administration. Also attending was well-known CBS radio correspondent Edward R. Murrow, the brother of Lacey Murrow.

Governor Martin conducted the ribbon-cutting ceremonies, followed by National Guard soldiers firing a 19-gun salute. Then the governor, his wife, Senator Homer Bone, and Representative John Coffee climbed into a 1923 Lincoln Touring Car loaned for the occasion by an Olympia auto dealership, the Titus Will Ford Company. The governor paid the first toll and motored across the bridge with American flags flying. In Gig Harbor and at McChord Field, marching bands, floats, and parades also celebrated with great fanfare.

"Everyone marveled," historian Murray Morgan wrote later, "at the gossamer grace of a structure so long." The 5,939-foot-long bridge was only 39 feet wide. The 2,800-foot central span between the two towers extended more than a half-mile. It was the third-longest suspension bridge in the world. Only San Francisco's Golden Gate Bridge (completed in 1937 with a center span of 4,200 feet) and New York's George Washington Bridge (completed in 1931 with a center span of 3,500 feet) were longer.

The streamlined ferry *Kalakala*, launched in 1935, also participated in the opening festivities. The *Kalakala* had the honor of making the last ferry run across the Narrows on July 2, 1940. More than 1,400 people took the ferry between Point Defiance and Gig Harbor, celebrating for hours with music and dancing sponsored by the Young Men's Business Club of Tacoma.[8]

ECONOMIC IMPACT

Civic leaders on both sides of the Sound hailed the positive economic impacts of the bridge project. During the 19 months of construction, wages paid to workers totaled $1,142,638. That million-dollar-plus payroll pumped much needed cash into the hands of local merchants struggling to survive through the Great Depression. Steady sales sustained many grocery and drug stores, furniture outlets, and other businesses, and stimulated the real estate market. The locality's population began to grow from the time construction started. The bridge boosted the local economy by an estimated $11 million—ten times the salaries figure.[9]

On the Peninsula, completion of the bridge led directly to land development for new homes and businesses. Agricultural acreage also increased because farmers could more easily deliver produce to the larger urban markets in Tacoma and Seattle. Timber companies now had ready access to the wealth of trees on the Olympic Peninsula, and some mineral resources could be more easily exploited with the quicker route to Tacoma's industrial centers. Residents and tourists avidly began to enjoy new recreational opportunities. The bridge also established a closer link between the state's two federal parks, Olympic National Park (only recently established in 1939) and Mount Rainier National Park.

Local journalists hailed the positive effects for Peninsula residents. One declared, "Everywhere you turn from the time you reach the west end of the new span, you sense the throb of progress that has hit the district as a result of construction of the bridge." A large new service station now stood beside the highway to greet motorists, while at a nearby harbor entrance, five ferries sat idly tied to the docks.

TRAFFIC AND TOLLS

User tolls on the 1940 Narrows Bridge, of course, were necessary for repaying the RFC's loan. The tolls were controversial, however, because they were more expensive than for the old ferry service. A car and driver paid 75¢ each way ($1.50 round trip); pedestrians were charged 15¢. Despite the complaints, traffic quickly far surpassed all expectations. Indeed, local residents and the Toll Bridge Authority had long thought that the new bridge would bring a population boom and economic boost, but actual use of the bridge dramatically exceeded their highest hopes. On opening day, more than 2,000 cars crossed the bridge.

The toll plaza and booths, looking west, July 1940.
Washington State Archives

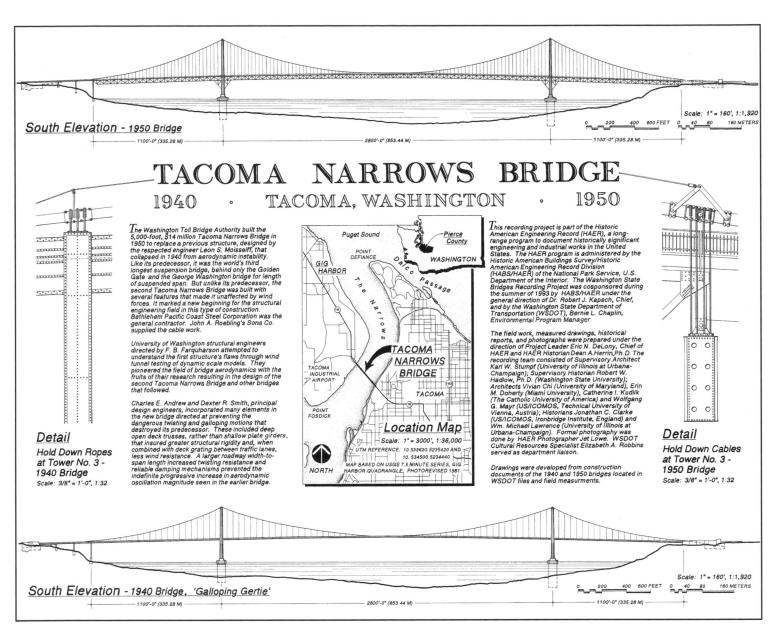

South Elevation - 1950 Bridge

Scale: 1" = 160', 1:1,920

1100'-0" (335.28 M) · 2800'-0" (853.44 M) · 1100'-0" (335.28 M)

0   200   400   600 FEET   0   40   80   160 METERS

TACOMA NARROWS BRIDGE
1940 · TACOMA, WASHINGTON · 1950

The Washington Toll Bridge Authority built the 5,000-foot, $14 million Tacoma Narrows Bridge in 1950 to replace a previous structure, designed by the respected engineer Leon S. Moisseiff, that collapsed in 1940 from aerodynamic instability. Like its predecessor, it was the world's third longest suspension bridge, behind only the Golden Gate and the George Washington bridge for length of suspended span. But unlike its predecessor, the second Tacoma Narrows Bridge was built with several features that made it unaffected by wind forces. It marked a new beginning for the structural engineering field in this type of construction. Bethlehem Pacific Coast Steel Corporation was the general contractor. John A. Roebling's Sons Co. supplied the cable work.

University of Washington structural engineers directed by F. B. Farquharson attempted to understand the first structure's flaws through wind tunnel testing of dynamic scale models. They pioneered the field of bridge aerodynamics with the fruits of their research resulting in the design of the second Tacoma Narrows Bridge and other bridges that followed.

Charles E. Andrew and Dexter R. Smith, principal design engineers, incorporated many elements in the new bridge directed at preventing the dangerous twisting and galloping motions that destroyed its predecessor. These included deep open deck trusses, rather than shallow plate girders, that insured greater structural rigidity and, when combined with deck grating between traffic lanes, less wind resistance. A larger roadway width-to-span length increased twisting resistance and reliable damping mechanisms prevented the indefinite progressive increase in aerodynamic oscillation magnitude seen in the earlier bridge.

Location Map
Scale: 1" = 3000', 1:36,000

UTM REFERENCE: 10.533620.5235420 AND 10. 534500.5234440
MAP BASED ON USGS 7.5 MINUTE SERIES, GIG HARBOR QUADRANGLE, PHOTOREVISED 1981.

NORTH

Puget Sound · *Pierce County* · WASHINGTON

POINT DEFIANCE · GIG HARBOR · Dalco Passage

The Narrows

TACOMA NARROWS BRIDGE

TACOMA INDUSTRIAL AIRPORT · TACOMA

POINT FOSDICK

This recording project is part of the Historic American Engineering Record (HAER), a long-range program to document historically significant engineering and industrial works in the United States. The HAER program is administered by the Historic American Buildings Survey/Historic American Engineering Record Division (HABS/HAER) of the National Park Service, U.S. Department of the Interior. The Washington State Bridges Recording Project was cosponsored during the summer of 1993 by HABS/HAER under the general direction of Dr. Robert J. Kapsch, Chief, and by the Washington State Department of Transportation (WSDOT), Bernie L. Chaplin, Environmental Program Manager.

The field work, measured drawings, historical reports, and photographs were prepared under the direction of Project Leader Eric N. DeLony, Chief of HAER and HAER Historian Dean A.Herrin,Ph.D. The recording team consisted of Supervisory Architect Karl W. Stumpf (University of Illinois at Urbana-Champaign); Supervisory Historian Robert W. Hadlow, Ph.D. (Washington State University); Architects Vivian Chi (University of Maryland), Erin M. Doherty (Miami University), Catherine I. Kudlik (The Catholic University of America) and Wolfgang G. Mayr (US/ICOMOS, Technical University of Vienna, Austria); Historians Jonathan C. Clarke (US/ICOMOS, Ironbridge Institute, England) and Wm. Michael Lawrence (University of Illinois at Urbana-Champaign). Formal photography was done by HAER Photographer Jet Lowe. WSDOT Cultural Resources Specialist Elizabeth A. Robbins served as department liaison.

Drawings were developed from construction documents of the 1940 and 1950 bridges located in WSDOT files and field measurments.

Detail
Hold Down Ropes at Tower No. 3 - 1940 Bridge
Scale: 3/8" = 1'-0", 1:32

Detail
Hold Down Cables at Tower No. 3 - 1950 Bridge
Scale: 3/8" = 1'-0", 1:32

South Elevation - 1940 Bridge, 'Galloping Gertie'

Scale: 1" = 160', 1:1,920

1100'-0" (335.28 M) · 2800'-0" (853.44 M) · 1100'-0" (335.28 M)

0   200   400   600 FEET   0   40   80   160 METERS

Karl W. Stump and Wolfgang G. Mayr, HAER

Traffic soared to more than triple the figures projected by surveys. In 1939, the ferry service had carried 205,842 vehicles across the Narrows, an average of 564 per day. In the first four months after the Narrows Bridge opened (July to October 1940), some 2,044 vehicles a day paid the toll and crossed. The four-month total of 265,748 vehicles far exceeded the number of cars that crossed on the ferry system during the entire year of 1939.

Because traffic—and, therefore, revenue—exceeded all projections, the WTBA soon reduced the commuter toll, and two months later refinanced the bond issues and reduced the basic toll by another 5¢. By October, the fare for a car and driver stood at 50¢ each way.[10]

"Galloping Gertie" Earns Her Name

Even in a light breeze, the Narrows Bridge moved. Suspension bridges were supposed to move, but this was different. In the first weeks after the opening, newspaper accounts referred to this as "the bounce," "the buck," or "the

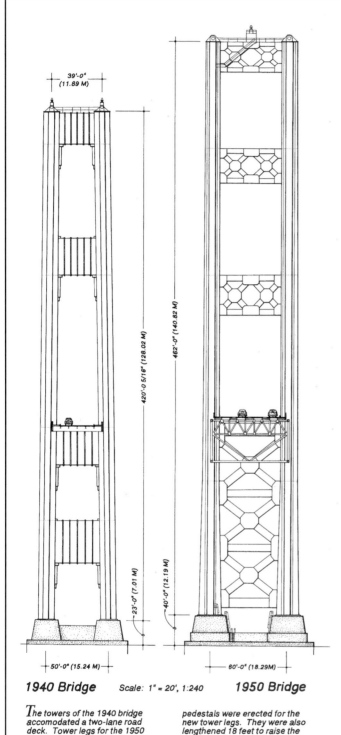

39'-0"
(11.89 M)

420'-0 5/16" (128.02 M)

462'-0" (140.82 M)

23'-0" (7.01 M)

40'-0" (12.19 M)

50'-0" (15.24 M)

60'-0" (18.29M)

1940 Bridge   Scale: 1" = 20', 1:240   **1950 Bridge**

The towers of the 1940 bridge accomodated a two-lane road deck. Tower legs for the 1950 bridge were designed for a four-lane road deck. Wider pier pedestals were erected for the new tower legs. They were also lengthened 18 feet to raise the tower steel above the Narrows' corrosive salt water.

Tower Elevations

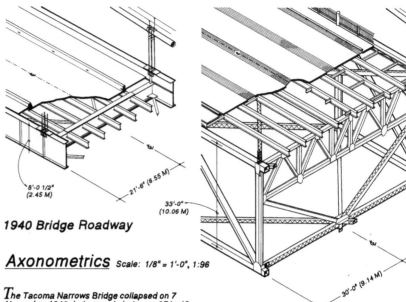

8'-0 1/2"
(2.45 M)

21'-6" (6.55 M)

1940 Bridge Roadway

33'-0"
(10.06 M)

30'-0" (9.14 M)

1950 Bridge Roadway

Axonometrics   Scale: 1/8" = 1'-0", 1:96

The Tacoma Narrows Bridge collapsed on 7 November 1940 during a gale between 35 to 42 miles per hour, with a wind pressure of only five pounds per square foot. The steady wind's effects on the structure produced a fluctuating resultant force that synchronized in timing and direction with the bridge's natural harmonic motions (figs. 1&2), progressively amplifying them to destructive levels. Both vertical and torsional oscillations contributed to the failure of the bridge. The bridge's inherent weakness and susceptibility to these winds lay in its shallow stiffening girders and its narrow roadway.

Theodore von Karman, who had pioneered wind tunnel analysis at the California Institute of Technology, argued that the bridge deck's aerodynamic shape was a more important factor in its failure than its lightness and flexibility. Von Karman suspected that the bridge had experienced vortex shedding, a condition where objects like airplane wings or bridge decks displace air flowing around them and form eddies or vortices, which may induce vibration in the object (figs 3&4). He believed that wind flowing over the bridge's solid girder side plates created shedding that when

combined with the flutter and resonance already present in the deck produced the violent oscillations that caused the catastrophic failure (figs. 5&6).

Designing the replacement bridge's deck stiffening system involved subjecting dynamic scale models to wind tunnel testing to better understand wind effects on them.

Designers for the 1950 bridge were not satisfied with their ability to eliminate torsional and vertical movements in their proposed structure. They hoped to enhance their design's natural damping ability with mechanical devices. One of these was a double-lateral bracing system in the stiffening truss. It increased torsional frequency motion and tortional stiffness.

Damping Mechanism
1950 Bridge
Scale: 1/16" = 1'-0", 1:192

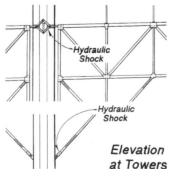

Hydraulic Shock

Hydraulic Shock

Elevation at Towers

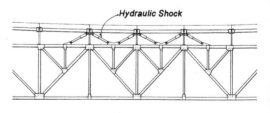

Hydraulic Shock

Elevation at Mid-Span

To eliminate torsional and vertical movement cylindrical hydraulic shock absorbers were used at three points on the bridge: coupling the top of the stiffening truss at mid-span with the suspension cables, connecting between the top chords of the main span and side span stiffening trusses, and extending as outriggers from the trusses' bottom chords to the towers.

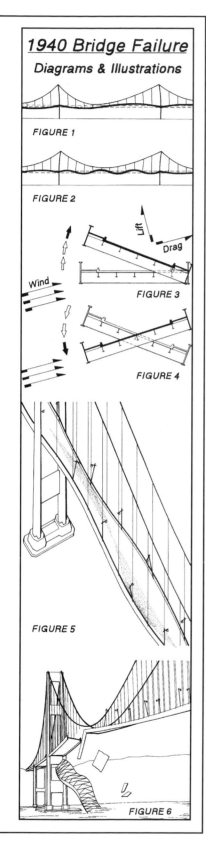

1940 Bridge Failure
Diagrams & Illustrations

FIGURE 1

FIGURE 2

Lift

Drag

Wind

FIGURE 3

FIGURE 4

FIGURE 5

FIGURE 6

Wolfgang G. Mayr, HAER

ripple." A variety of locals, engineers, and other observers described the movement in various other ways, too—gallop, wave, undulation, up and down, breathing, crests and troughs, peaks and valleys, rising and falling, roller coaster, vertical oscillation, and vertical flexibility.

The roadway sometimes bounced in a wind of only three or four mph. Often, several waves of two to three feet (and on a few occasions up to five feet) would roll from one end of the center span to the other. There seemed to be no correlation between wind speed and the size of the waves. Sometimes the half-mile-long center span would "bounce" for a few moments then stop. Other times the waves lasted for six or even eight hours.

Thrill-seekers drove from miles around when the ripples started, while other motorists who felt motion sickness avoided the bridge. But for adventurous spirits, the bridge was an amusement ride. At times drivers crossing the span saw a car in front of them suddenly disappear when a section of the roadway dropped. Moments later, the vehicle reappeared as the road rose. According to one report, on a couple of occasions drivers experienced waves 10 feet high.

Howard Clifford, then a young reporter for the *Tacoma News Tribune*, often crossed the bridge. "It was an odd sensation," Clifford recalled. "You'd be driving along and there'd be a car forty or fifty yards ahead of you, and it would disappear in the valley of the wave. And, then you'd come up on the crest and see the car again. It was a very weird sensation."

Another area resident who experienced the "fun" of the bridge's peculiar motions was Beverlee Storkman, then a sophomore at the College of Puget Sound in Tacoma. "One time, when we were coming back across the bridge in August 1940," Storkman recalled, "the traffic was so slow that Dad let us out of the car to walk a little. We took a few steps and got the feel of it—it was just accepted that the bridge rolled."[11]

No one is certain how "Galloping Gertie" got her name. Most likely, it came from the bridge workers. Many of the most experienced workmen on the Tacoma Narrows project had followed bridge construction all over the country. Called "boomers," they formed the nucleus of most crews. Often, they came from families in which bridge building was something of a tradition, with skills and folklore passed from one generation to the next. Very possibly one of their grandfathers had worked on the Wheeling Bridge. The name "Galloping Gertie" was first used for the Wheeling Bridge, according to bridge historian James Hopkins.[12] Charles Ellet Jr. built this 1,010-foot suspension span in 1849 over the Ohio River in Wheeling, West Virginia. Back then, it was the longest suspension bridge in the world. Less than five years after completion, it collapsed in a windstorm in May 1854.

Many "boomers" came to the Narrows project from the newly completed Golden Gate span in San Francisco. Interestingly, the Golden Gate Bridge also had a tendency to bounce in high winds when first finished, although much less so than the Tacoma Narrows Bridge. In early May 1940, as crews were building forms and laying concrete for the roadway, and the Narrows Bridge began to ripple, the workers, probably boomers, dubbed it "Galloping Gertie." Local residents picked up the nickname and it stuck. Only after the collapse in November 1940 did the name appear in the newspapers.

Notes

1. "Huge Narrows Span Collapses into Sound," *Seattle Post-Intelligencer*, November 8, 1940.

2. Murray and Rosa Morgan, *South on the Sound: An Illustrated History of Tacoma and Pierce County* (Woodland Hills, California: Windsor, 1984), 115–16; "Fight for Narrows Toll Span Recalled," *Tacoma Ledger*, October 3, 1938. The WTBA minutes are contained in Box 5, and the applications to the PWA are in Box 56, Bridge Files, Washington Toll Bridge Authority, Washington State Archives (hereafter cited as WTBA, WSA).

3. "Tacoma Narrows Bridge, Final Report on Design and Construction," by Clark Eldridge, 1940, 9, Box 43, WTBA, WSA; "Clark H. Eldridge, An Autobiography, 1896–1982," 9, a manuscript copy loaned to the author by Eldridge's son, C.W. Eldridge. Though Eldridge was the effective head of the Narrows project, much of the correspondence with Moisseiff and other parties relating to design and construction was sent under Murrow's name, followed with the words, "By: Clark H. Eldridge, Bridge Engineer."

4. Letter, L.V. Murrow (WTBA) to L.R. Durkee (PWA), March 30, 1940, Box 55, WTBA, WSA; Eldridge quoted, "Row over Bridge's Collapse," *Tacoma News Tribune*, November 11, 1940; letter (copy), Eldridge to Morton McCartney (RFC), November 9, 1940, "Tacoma Narrows Bridge, Progress Reports 1940," Box 102, Theodore von Karman Papers, California Institute of Technology Archives, Pasadena, California (hereafter cited as von Karman Papers, Caltech).

5. "Bridge Grant Climax of Long Hard Drive," *Tacoma News Tribune*, June 28, 1938 (interestingly, this issue includes David Steinman's 1929 proposal sketch of a Narrows bridge); "Tacoma Narrows Bridge, Dream of Man for More Than 20 Years, Comes True," *Tacoma Times*, July 1, 1940; "Exciting Days Began Back on Nov. 22, '38," *Tacoma News Tribune*, July 1, 1940.

6. Telegram, Murrow to Eldridge, July 8, 1938, letters, Eldridge to Moisseiff and Moran & Proctor, July 9, 1938; letters, Moisseiff to Murrow, July 18 and 27, 1938, Box 42, WTBA, WSA (Moisseiff's letters to Murrow are reprinted in Advisory Board, *Failure of the Tacoma Narrows Bridge*, II, 1–8); "Row over Bridge's Collapse," *Tacoma News Tribune*, November 11, 1940.

7. "Report of Board of Consulting Engineers, The Tacoma Narrows Bridge," August 31, 1938, and letter, Leon Moisseiff to WTBA, May 27, 1939, Box 42, WTBA, WSA. Signatures on the August 31 report included Consulting Engineers Board members Charles E. Andrew, Chairman, Luther E. Gregory, and Ray B. McMinn. Moisseiff's telegram and follow-up letter to Murrow, September 24, 1938, are included in Box 42, WTBA, WSA. At the time, Moisseiff had been on retainer for more than a year for the Aluminum Company of America in regard to applying aluminum in bridge designs.

8. Morgan, *South on the Sound*, 116; "Bridges, Airbase Dedication Ready," *Seattle Post-Intelligencer*, July 1, 1940; "Official Opening: Tacoma Narrows Bridge and McChord Field, June 30–July 4, 1940, A.D." dedication program, Box 42, WTBA, WSA; "First Autos Cross Narrows," *Seattle Times*, July 2, 1940; Steven J. Russell, *Kalakala: Magnificent Vision Recaptured* (Seattle: Puget Sound Press, 2002). The 1923 Lincoln that led the parade later was warehoused for more than fifty years. A complete refurbishing in 1996–1999 restored it for use in parades. The Titus Will Ford Company (family descendants of the original owner, Leon Titus) owns the vehicle. Greg Anderson, e-mail to author, September 13, 2005.

9. Allan J. Locke, "The Development and Economic Significance of the Tacoma Narrows Bridge, 1923–1953," M.A. Thesis, College of Puget Sound, 1956, 3–60; David L. Glenn, "$1,142,638 Total of Narrows Span Wages," *Tacoma Ledger*, August 11, 1940.

10. "Tacoma Narrows Bridge Tolls Reduced," *Engineering News-Record* 125 (August 8, 1940): 197.

11. Howard Clifford quote from the 30-minute documentary "Gertie Gallops Again," Tacoma Municipal Television, 1998; Beverlee Storkman, telephone interview, September 17, 2006.

12. Henry James Hopkins, *A Span of Bridges: An Illustrated History* (New York: Praeger, 1970), 231–32. Unfortunately, Hopkins does not cite the source of this information. Another writer, Dr. Tadaki Kawada, in an unpublished manuscript, suggests that the name "Galloping Gertie" derived from a story about a girl who danced herself to death in red ballet shoes. The reference is to "The Red Shoes," by Hans Christian Anderson, published in 1845. However, the name of the girl in Anderson's story was Karen.

Chapter Three

The 1940 Tacoma Narrows Bridge Machine

The 1940 Narrows Bridge, looking west from the Tacoma side.
WSDOT

Suspension Bridge Design to 1940

We live in an era of great suspension bridges. They have been built for centuries, but only became common in modern times. The earliest forms used ropes for cables. A major advance came in 1826 with the completion of the Menai Strait Bridge in Great Britain. It featured the world's longest single suspension span and represented a major advance by adopting linked iron bars, or "eyebars,"

for cables. Other builders soon started using wrought-iron wire for main cables.

The first use of steel-wire cables for a suspension bridge occurred in 1883 when workers completed the Brooklyn Bridge. Steel is extremely strong and ideally suited for cables. The Brooklyn Bridge designer was America's most famous suspension bridge engineer, John Roebling. Along with his son, Roebling developed methods of spinning wire cables that are still used today. Roebling also borrowed the foundation construction process developed by James Eads, who in 1874 pioneered the use of caissons when building the Eads Bridge in St. Louis.

Suspension bridge failures occurred periodically in the 19th century. At least ten—three in the United States and seven in Europe—either collapsed in windstorms or otherwise suffered significant damage in the seven decades between 1818 and 1889. Before the completion of the 1883 Brooklyn Bridge, people considered suspension bridges risky and unreliable. Furthermore, in an era when railroads dominated transportation in America, other types (such as the steel cantilever) dominated bridge architecture because of their ability to carry the weight of heavy trains.

After 1909, that began to change. The Manhattan Bridge in New York City introduced the "Deflection Theory" in suspension bridge design to the United States. One of the lead designers was Leon Moisseiff. The Deflection Theory had been formulated in Austria for concrete arch bridges before Moisseiff developed and applied it to suspension bridges.

A series of mathematical formulas were expressed in the theory. The central principle held that rigid, heavy, stiffening trusses were unnecessary. That was because the theory could mathematically—and accurately—describe how a suspended structure and its main cables interacted, or "deflected," under the stress of moving traffic. The cables themselves helped to stiffen the bridge, thus reducing vertical deflections, so the suspended structure needed only to distribute its own weight and traffic loads to the cables. Now engineers could, with mathematical precision, design a lightweight suspended structure to flex, so long as bridge users did not notice its movements.

In the early 1930s, Moisseiff and an associate, Frederick Lienhard, elaborated on the theory to account for the stresses from winds striking laterally (from the side). The main cables helped restrain lateral movement in a suspended structure so it did not have to resist the wind on its own. As a result, engineers now could design even narrower suspended structures, an important factor for the Depression-era bridges that only required two lanes of traffic for automobiles.[1]

The application of these theories had a profound impact. By the late 1920s and into the 1930s, long-span suspension bridges were being built across the country. Engineers designed narrower, lighter, and more slender and flexible spans that could limit vertical deflections from traffic loads and horizontal deflections from wind forces. This opened the door to the adoption of more flexible and inexpensive plate girders in the suspended part of the structures. The amount of steel needed to build suspension bridges could be hundreds of tons less than what was required in previous designs, greatly reducing costs. These structures were theoretically sound, cheaper to build, and considered beautiful by many.

PREDECESSOR BRIDGES—NARROWER, LIGHTER, MORE FLEXIBLE

The 1931 George Washington Bridge, designed by Othmar Ammann with assistance from Leon Moisseiff, dramatically bolstered confidence in the progressive evolution of suspension bridge design. In 1931, it was the longest suspension span in the world. The bridge had a relatively flexible suspended structure, with little stiffening against vertical movement. Its bare steel towers ended the use of masonry coverings. Engineers and the public applauded the span as "a vision of simplicity and grace." Afterward, Ammann, Moisseiff, and David Steinman designed bridges with virtually no stiffening.

However, aerodynamic problems began to appear as this trend accelerated in the late 1930s. The 1937 Golden Gate Bridge, to which Moisseiff also contributed, was noted for its exceptional narrowness and flexibility, but this introduced a new level of vulnerability to wind forces that soon became apparent

The 1940 Tacoma Narrows Bridge and Its Contemporaries

	George Washington	Golden Gate	Bronx-Whitestone	Tacoma Narrows
Year completed	1931	1937	1939	1940
Cost	$59.5 million	$35 million	$19.7 million	$6.4 million
Length of center span	3,500 ft	4,200 ft	2,300 ft	2,800 ft
Girder (stiffening member) depth	29 ft	25 ft	11 ft	8 ft
Width (of wind truss)	106 ft	90 ft	74 ft	39 ft
Ratio: girder depth to length of center span	1:120	1:168	1:209	1:350
Ratio: width to length of center span	1:33	1:47	1:31	1:72

to engineers. The 1939 Bronx-Whitestone Bridge was the first suspension structure to have steel towers without diagonal cross-bracing and to use a solid plate girder for the roadway deck. Experts hailed it as "the ultimate" in suspension design. However, vertical oscillations (waves and vibrations) of the roadway began appearing. This also was apparent in David Steinman's 1939 Deer Isle Bridge.

The 1940 Narrows Bridge represented a culmination of the trend to build longer bridges with a narrower road width and a minimum of stiffening. The design emphasized economy, lightness, slimness, and flexibility. The span's width-to-length ratio was 1 to 72—a dramatic increase over the previous record-holder, the Golden Gate Bridge, at 1 to 47. The Tacoma Bridge also was the cheapest.

Engineering Challenges in 1938

Engineers faced several significant tests in constructing the Narrows Bridge. First and foremost was the geography of the Narrows itself. The depth of the water meant that the piers would be some of the deepest ever constructed, almost double that of the Golden Gate's piers. The major difficulty occurred in the early part of construction. During their floating stages, the caissons had to be anchored in treacherous tides moving at 12.5 feet per second.

The length of the crossing also posed difficulties, and traffic surveys had added to the challenge. Studies expected modest use of the span, thus there was no way to justify more than two lanes. Consequently, circumstances called for a structure that was both long and narrow.

Eldridge's Design

Clark Eldridge, the State Highway Department's lead engineer on the project, developed the bridge's original design from late 1937 to the spring of 1938. He envisioned a center span of 2,600 feet, two side spans of 1,300 feet each, two sidewalks each four feet wide, trusses and cables 39-feet center-to-center, and stiffening trusses 25 feet deep. Because the approaches differed in height (the Tacoma end was almost 20 feet higher), Eldridge designed towers of different heights—the east tower at 476 feet 6 inches, and the west tower at 463 feet 6 inches. Each tower had five bracing struts between the legs above the deck and

Eldridge's design; elevation detail, May 23, 1938.
Washington State Archives

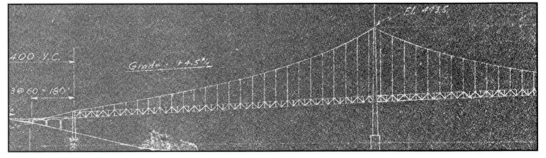

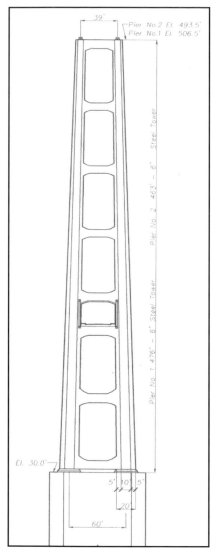

Eldridge's original tower design.
2003 replication, Laurie Carron, architect

three below. Eldridge's designed called for portal struts, unlike the cross-bracing used in the San Francisco-Oakland Bay Bridge towers.[2]

The design also included a "through truss" without upper lateral supports, an approach typical of the period. Eldridge planned a bridge with a center span-to-width ratio of 1 to 66. Although this represented a narrower span than the 1 to 47 ratio of the Golden Gate Bridge, Eldridge's 25-foot deep truss would give substantial stiffness and weight to the structure.

However, Eldridge's design was never built. The Public Works Administration's requirement that the Washington Toll Bridge Authority hire Leon Moisseiff and the firm Moran & Proctor resulted in critical changes. Their final design substantially replaced Clark Eldridge's plan.

MOISSEIFF'S DESIGN

The most evident of Leon Moisseiff's changes pertained to the superstructure. He used Eldridge's same basic specifications for the roadway width, but in every other respect designed a very different superstructure.

Moisseiff criticized Eldridge's effort on aesthetic and economic grounds, not for engineering concerns. His report to Lacey Murrow in late July 1938 declared, "Unless there are very valid reasons which compel the making of the towers of unequal heights[,] the towers should be of identical design and fabrication. Economic fabrication and good appearance demand it. The symmetry of the structure should be adhered to." Eldridge's 25-foot-deep stiffening truss, said Moisseiff, imposed a "great cost." The eight-foot solid plate girder that he substituted would be cheaper, and would yield a "neat and pleasing appearance."[3]

Moisseiff's side spans were 1,100 feet long, and the center span was 2,800 feet in length. He changed the towers to the same height, 425 feet above the piers, and also reduced the number of tower bracing struts, both above and below the roadway. He substituted two

strut panels to connect the tower legs above the deck and two below.

Indeed, Moisseiff's Modernist-Art Deco design for the towers and anchorages set a new standard for suspension bridge aesthetics. However, this also would result in a much lighter, more flexible bridge, with a center span-to-width ratio of 1 to 72. It was unprecedented, substantially beyond the Golden Gate Bridge's 1 to 47 ratio. Ironically, the Narrows Bridge would be the first that Moisseiff largely designed by himself. Although he had participated in building America's long suspension bridges for nearly four decades, he had never been a primary, leading designer.

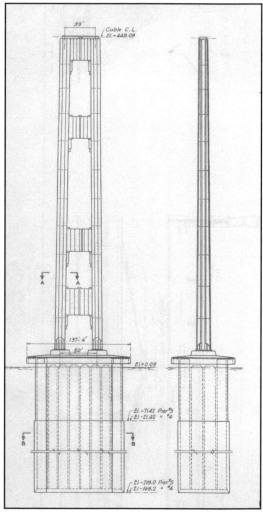

Moisseiff's design; tower face and side views, 1938.
Washington State Archives

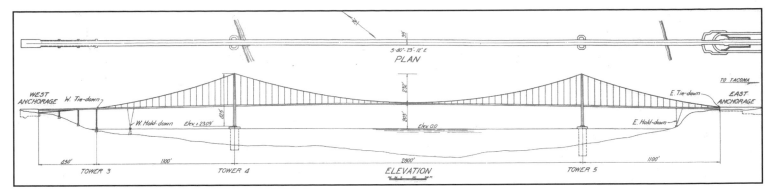

PLAN

WEST ANCHORAGE   W. Tie-down   TO TACOMA   E. Tie-down   EAST ANCHORAGE

W. Hold-down   Elev. +230.9'   Elev. 0.0'   E. Hold-down

450'   1100'   2800'   1100'

TOWER 3   TOWER 4   ELEVATION   TOWER 5

Complaints from contractors presented the next hurdle. When the substructure plans developed by Moran & Proctor went out for bid, contractors protested to the State Highway Department. The pier design, they said, would be impossible to build. However, contractors agreed with Clark Eldridge's original plans, and the PWA approved modifications in the bid. Thus, the bridge was constructed following Eldridge's plan for the piers and Moisseiff's design for the superstructure.

Starting Construction

A combination of companies held contracts—the Pacific Bridge Company of San Francisco (general contractor), the General Construction Company of Seattle, and the Columbia Construction Company; these firms built the piers and west anchorage. The Bethlehem Steel Company of Pennsylvania held a sub-contract for the steel superstructure. John A. Roebling's Sons Company of New York produced the cable wire, and Woodworth and Cornell Company was the subcontractor for the east anchorage and roadway.

Clark Eldridge and his staff prepared detailed design drawings—39 sheets; dated August 6 and September 7, 1938—in accordance with Moisseiff's general directions and mathematical calculations. Construction started in November 1938, and workmen would complete the bridge quickly, in only 19 months, setting a world record. Typical construction time for such a project previously

was twice that long—at least three or four years.[4]

Project Milestones

Work began with land clearing at the approaches on November 23, 1938, and the placement of caisson anchors in the channel beginning January 25, 1939. By March 18, tugboats towed the first caisson to the Narrows from the Seattle docks where it had been built. Meanwhile, excavation for the cable anchorages on either shoreline began and was completed in mid-May.

In the caisson work, the pier construction for the west tower was finished on July 13, 1939, and on September 15 for the east tower; construction of the towers themselves was completed between August and November 1939.

Moisseiff's design, 1938.
Washington State Archives

A caisson being towed to the bridge site, 1939.
WSDOT

Spinning towers, catwalks, and cables, January 1940.
WSDOT

Cable spinning began in January 1940 and ended March 5. By March 21, the first section of the steel plate girder of the suspended structure was put in place. Crews finished the steel floor system (girders, beams, and stringers) on May 6, at a rate of 200 feet per day.

Concrete work began immediately and was completed by June 10, at an average of 300 feet per day.

By the prescribed deadline of June 30, 1940, workmen finished the concrete roadway, sidewalks, and curbs. Painting the steel superstructure had started in mid-April and was largely finished by the end of June, but dragged on for several more months. On July 1, 1940, the Tacoma Narrows Bridge stood ready for the official opening ceremonies.

Anchors for the Caissons

In order to hold the caissons steady against the current, engineers and workmen distributed concrete anchors (measuring 12-feet by 12-feet by 51-feet 6-inches) in a 900-foot diameter circle around each caisson. The caissons were attached to the anchors with $1^9/_{16}$ inch diameter cables. Thirty-two anchors were required for the east pier (Pier #5) and 24 for the west pier (Pier #4) caissons. Each anchor weighed approximately 600 tons.

Beginning of a caisson platform, and a 600-ton cement anchor block being dumped from a barge, 1939.
WSDOT

Caissons and Tower Piers

Construction of the tower piers at near world-record depths proved to be the most difficult challenge of its kind ever attempted. Swift tides created a dangerous environment requiring creative engineering solutions. The tidal currents shifted constantly and caused a treacherous swirling action around the caissons. The east pier was the more difficult of the two. The caisson's cutting edge eventually touched bottom at 140 feet, then penetrated another 90 feet into the sea floor. Final depth of the east pier measured 224 feet below mean sea level.

The caissons, consisting of steel trussing and girders arranged within wooden sheathing, measured roughly 66 feet by 119 feet. They were built up to 36 feet in height off-site, then towed to the Narrows. At the base was a sharp metal cutting edge. Once tugboats positioned a caisson, workmen began pouring concrete in open cells. The caisson sank under its own weight until the top was near sea level. Then workers built a 12-foot-high wooden hull at the top, with a steel framework and partitions. Concrete pouring then continued.

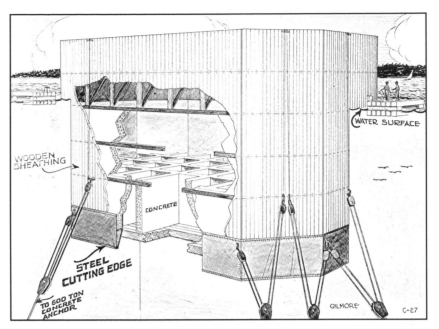

Caisson cutaway drawing, 1939.
Washington State Archives

The rapid tides made it difficult for the concrete construction barges to remain in position next to the caissons. Sometimes, in bracing against the tidal current, one end would be eight feet lower in the water than the other.

The three-month process was repeated in 12-foot segments until the cutting edge reached the floor of the Narrows. Then, workers removed boards in the caisson's bottom section and started excavation. Clamshell buckets scooped away compacted gravel, sand,

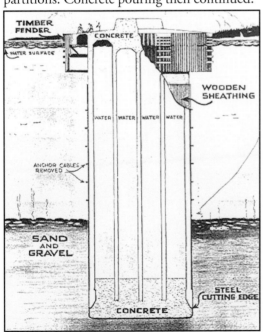

Pier cross-section drawing, 1939.
Washington State Archives

The west tower (#4) begins to rise on its pier (foundation).
Washington State Archives

and mud, lifting it up through the caisson's hollow concrete cells to the surface. Slowly, the caisson sank to the desired depth for a solid pier foundation. Finally, workers placed a concrete cap on each pier.

Anchorages

Meanwhile, construction of the anchorages on the opposite shorelines proceeded. Beginning in March 1939, workers removed tons of dirt, sand, gravel, and rock on the bluffs to prepare the anchorages and the highway approaches. They then poured tons of concrete for the anchorages. Next came placement of the eye-bars. By the third week of May, completion of the anchorages was largely suspended until after the spinning of the cables.

Towers

The 425-foot tall towers (measured from the top of each pier to the center of the cables at

West anchorage with completed south cable, March 5, 1940.
WSDOT

Cable spinning machine at east anchorage, November 3, 1939.
WSDOT

the top of each tower, or, as more commonly measured, 448 feet from mean sea level) consisted of carbon steel plates (mostly ½ inch). Workmen riveted the plates into place.

Engineers designed the towers to deflect (bend) toward shore three feet during construction, and three inches in their final dead-load positions. The tower legs were built as cells. Longitudinally (parallel to the bridge's deck), each leg tapered from 19 feet at the base to 13 feet at the top. Transversely (90 degrees to the deck), the leg cells measured 13 feet, from bottom to top. The legs of each tower were spaced 50 feet apart (center-to-center) at the base, and 39 feet apart (center-to-center) at the top. At the top of the towers was another important innovation; this was the first suspension bridge to have all-welded cable saddles, rather than cast steel as used previously.

Once the towers were completed, workers built catwalks. Next came the spinning of the main cables, followed by the addition of the plate girder suspended structure. Suspender cables would attach the main cables to the deck at 50-foot intervals with zinc connectors called "jewels."

West tower construction, August 20, 1939.
Washington State Archives

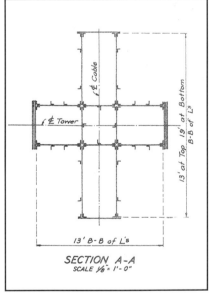

SECTION A-A
SCALE ⅛" = 1'-0"

Tower leg cross-section drawing, 1939.
WSDOT

Spinning towers and catwalks as seen from the east anchorage, March 22, 1940.
WSDOT

spinning equipment were installed. Then workmen hoisted a wire to the top of one tower. They repeated the process for a second wire. The two wires then were joined together to make a "traveler rope."

At each anchorage, the traveler rope was fitted with a "spinning wheel" and looped over a pulley. One end of wire from a spool was secured to an eye-bar at the anchorage and looped around the spinning wheel. The spinning wheel followed the traveler rope up over the towers at the cable saddle and across to the anchorage on the opposite shore.

When the desired number of wires had been accumulated, workers collected them into a strand. They then compacted the strands, and temporarily wrapped them with wire bands at intervals, before the final completion of the spinning.

Suspender Cables and Cable Bands

Vertical suspender cables (also called "suspenders," "hangers," or "hanger ropes") made of twisted wire were placed at 50-foot intervals

Cable spinning drawing, 1939.
Washington State Archives

Cable Spinning

Giant reels of steel wire waited at the east anchorage. The wire was the thickness of a pencil. Before the main cables could be constructed, the catwalks that would serve as platforms for workers spinning the cables had to be built. A boat took the first catwalk wire over the water. Once the catwalk was raised and finished, wires and supports for the cable

At the start of cable spinning, January 10, 1940. Lacey Murrow, WTBA engineers, and local politicans.
Washington State Archives

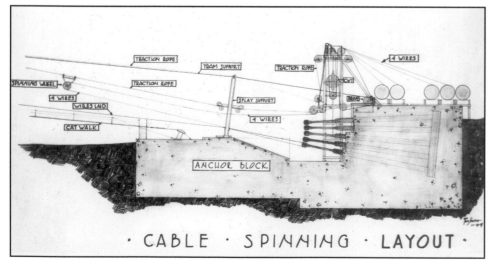

· CABLE · SPINNING · LAYOUT ·

East anchorage, June 28, 1939.
WSDOT

Cable splay to strand shoes, east
anchorage, March 1940.
Washington State Archives

Cable compacting,
March 5, 1940.
WSDOT

Cable wrapping machine, March 23, 1940.
Washington State Archives

Caulking cable band with lead wool, May 1940.
Washington State Archives

along the main cables for attachment to the road deck. Workmen looped the suspenders over the main cable with cable bands.

Next, they wrapped the main cables with steel wire for protection against the elements. Workmen hammered lead wool into the joints and the openings around the cable bands to make a watertight fit.

Suspended Structure

Pre-formed sections were used in constructing the plate girder suspended structure. Each was composed of two stiffening girders 39 feet apart, transverse (crossing) floor beams (each 25 feet long and 52 inches deep), and five lines of 21-inch "stringers" (lateral connectors running longitudinally), plus eventually the concrete roadway and sidewalk slabs. Both the 5¼-inch thick road slab and the 1½-inch sidewalk slab would be heavily reinforced with steel rods.

Section of the suspended structure, 1939.
WSDOT

39'-0"

Cable

Stiffening
Girder

26'-0" Roadway

Stringer

Lateral

Floorbeam

TYPICAL CROSS SECTION THRU BRIDGE

1½"∅ Galv. stl bridge strand
Ult strength 270,000#

2-¾" tie welded to cable bands

LONGITUDINAL SECTION AT CENTER MAIN SPAN

Riveting the suspended structure on the east-side span, May 1940.
WSDOT

First section of the suspended structure, 111-feet long,
being raised into place, March 21, 1940.
WSDOT

The plate girders were stiffened horizontally and vertically. Each one was a "web" plate 96-inches (8 feet) by ½-inch thick, with four ½-inch angles 8 inches by 6 inches, and two ½-inch covers measuring 20 inches. The girder and deck sections were pre-assembled in Seattle. Scows carried the sections to the construction site, then cranes with Chicago booms hoisted the sections into place and they were riveted. An interesting example of on-site ingenuity occurred when workers had trouble fitting the girders into place. The solution—the adjoining pieces were packed in dry ice, causing them to contract, then fitted together and riveted into place.

On November 13, 1939, at 7:45 A.M., the strongest earthquake in decades struck the region. The 6.2 magnitude jolt, with an epicenter some six miles south of Bremerton, shook the bridge. When engineers investigated, they were relieved to find no damage to the towers.

Finally, by April 1940 the final painting began. The bridge received three coats of "Narrows Green" paint. A newspaper article at the time called it "Chrome Green," noting how the color "gives the slim structure an air and blends happily with the sky, water and foliage of the surrounding bluffs."[5]

Painting the suspender cables, April 26, 1940.

Concrete work finished the roadway, June 1940. By this time workmen had dubbed the bridge "Galloping Gertie."
Washington State Archives

Notes

1. Henry Petroski, *Engineers of Dreams: Great Bridges Builders and the Spanning of America* (New York: Alfred A. Knopf, 1995), 270–72, 289–308; Richard Scott, *In the Wake of Tacoma: Suspension Bridges and the Quest for Aerodynamic Stability* (Reston, Virginia: ASCE Press, 2001), 7–40; F.B. Farquharson, *Aerodynamic Stability of Suspension Bridges with Special Reference to the Tacoma Narrows Bridge* (Seattle: University of Washington Press, 1954), 13–15; Leon S. Moisseiff, "Growth in Suspension Bridge Knowledge," *Engineering News-Record* 123 (August 17, 1939): 46–49; Advisory Board on the Investigation of Suspension Bridges, *The Failure of the Tacoma Narrows Bridge* (College Station, Texas: School of Engineering, Texas Engineering Experiment Station, 1944); Dino A. Morelli, "Some Contributions to the Theory of the Stiffened Suspension Bridge," Ph.D. Dissertation, California Institute of Technology, Pasadena, 1946. In addition to these sources, the author gratefully acknowledges suggestions from suspension bridge expert and author Richard Scott (cited above), and Tim Moore, Senior Structural Engineer for the Washington State Department of Transportation.

2. The Highway Department's application to the PWA, dated May 23, 1938, includes Clark Eldridge's preliminary design sketches (blueprint), Box 55, WTBA, WSA.

3. Letters, Moisseiff to Murrow, July 18 and 27, 1938, Box 20, WTBA, WSA; reprinted in Advisory Board, *Failure of the Tacoma Narrows Bridge*, II, 1–8.

4. The principal sources for the construction and features of the 1940 Tacoma Narrows Bridge include the following—"Tacoma Narrows Bridge, Tacoma Washington: Final Report on Design and Construction," by Clark H. Eldridge, 1940, Box 18, WTBA, WSA; "Tacoma Narrows Bridge, Tacoma Washington: Report on Construction of the Substructure," by H.F. Connelly, 1940, Box 18, "Galloping Gertie" Collection, WSDOT Records, Washington State Archives; "Tacoma Narrows Bridge, Tacoma Washington: Report on Construction of the Superstructure," by H.F. Connelly, 1940, Box 18, "Galloping Gertie" Collection, WSDOT Records, Washington State Archives; Charles Andrew, "General Report on the Design of the Tacoma Narrows Bridge," January 15, 1942, Box 53, "Galloping Gertie" Collection, WSDOT Records, Washington State Archives; James Bashford, "Unusual Problems Attend Narrows Bridge Construction," *Compressed Air Magazine*, December 1939, 6033–35; "Bridge Dedication Number Lake Washington Floating Bridge-July 2, Tacoma Narrows Suspension Bridge-July 1," *Pacific Builder and Engineer*, July 6, 1940; Clark H. Eldridge, "The Tacoma Narrows Suspension Bridge," *Pacific Builder and Engineer* 46 (July 6, 1940): 35–40; Clark H. Eldridge, "The Tacoma Narrows Bridge," *Civil Engineering* 10 (May 1940): 299–302; Bruce Johnston and H.J. Godfrey, "Test Model of the Tacoma Narrows Anchorage Bar," *Welding Journal* 18 (August 1939): 253–59; J. Jones, "Welded Cable Saddles for Tacoma Narrows Bridge," *Engineering News-Record* 123 (December 7, 1939): 91–92; Lacey V. Murrow, "Construction Starts on the Narrows Bridge," *Pacific Builder and Engineer* 45 (March 4, 1939): 34–37; Thomas W. Neill, "Two New Washington Bridges that Write History," *Washington Motorist* 21 (July 1940): 4–5, 12; "Official Opening: Tacoma Narrows Bridge and McChord Field, June 30–July 4, 1940, A.D." dedication program (Tacoma: Johnson-Cox Company), WSDOT records, "Galloping Gertie Collection," Washington State Archives; "Pacific Northwest Bridges Completed," *Engineering News-Record* 125 (July 11, 1940): 2; "Plywood Forms Are Used in Caissons for First Time on the Tacoma Narrows Bridge," *Pacific Builder and Engineer* 45 (May 6, 1939): 30–32; Fred K. Ross, "Big Caisson Landed in Deep Tide Rip," *Pacific Builder and Engineer* 45 (June 3, 1939): 35–37, 48.

5. "Bridge Color? Chrome Green," *Tacoma News Tribune*, July 1, 1940.

General

Cost	$6,618,138
Total structure length	5,939 feet
Suspension section	5,000 feet
Center span	2,800 feet
Shore suspension spans (2), each	1,100 feet
East approach and anchorage	345 feet
West approach and anchorage	594 feet

Suspended Structure

Roadway height above water (mean sea level)	190 feet at towers, 208 feet at center span
Center span vertical clearance above mean sea level)	203 feet
Weight of center span	5,700 lb./ft
Traffic lanes	2
Width between cables	39 feet
Width of sidewalks(2), each	5 feet
Width of roadway	26 feet
Thickness of roadway	5¼ inches, reinforced concrete
Suspender cables, intervals	50 feet
Number of girders and type	2 plate girders
Depth of stiffening girder	8 feet
Ratio, deck width to center span	1:72
Ratio, deck depth to center span	1:350

Anchorages

Weight of each anchorage	52,500 tons
Concrete in each anchorage	20,000 cu. yds.
Structural steel in both anchorages	592 tons
West anchorage (concrete anchor block and gallery)	169 feet long, 70 feet wide, 66 feet deep
East anchorage (concrete anchor block and gallery), approach, administration buildings, and toll house	173 feet long, 70 feet wide, 50 feet deep
West anchorage, construction and cost	Woodworth & Cornell $299,734
East anchorage, construction and cost	Pacific Bridge Co., $272,273
Cable anchor bars	38 in each anchorage, 54 feet long

Cables

Diameter of main suspension cable	17½ inches
Weight of main suspension cable	3,817 tons
Weight sustained by cables	11,250 tons
Sag ratio (vertical distance between tower top and cable elevation at mid-span, as a ratio of main span length)	1:12
Number of wire strands in each cable	19
Number of No. 6 wires in each strand	332
Number of No. 6 wires in each cable	6,308
Total length of wire, two cables	74,926,424 feet = 14,191 miles
Number of wires carried by spinning wheel per trip	8
Number of reels of wire	440
Length of wire per reel	32 miles

Length of wire in one strand	370 miles
Time needed by spinning wheel to cross Narrows per trip (one way)	10 minutes
Rate of travel by spinning wheel	700 feet per minute
Number of round trips per 24-hour period by spinning wheel	60 to 70
Cost of cables, bands, suspenders, fittings, etc.	$1,335,960

Towers

Height above mean sea level	448 feet
Height above piers	425 feet
Weight of each tower	1,927 tons
Height above roadway	230 feet
Legs (2 per tower) at base, each	19 feet 1½ inches by 13 feet 5½ inches
Legs (2 per tower) at top, each	13 feet by 13 feet, 39 feet apart
Time to construct, west tower (#4)	25 days (8 hours/day)
Time to construct, east tower (#5)	21 days (8 hours/day)
Rivets in each tower	52,500
Cost of both towers	$ 542,144

Piers and Caisson Anchors

West pier (#4), total height	198 feet
West pier (#4), depth of water	120 feet
West pier (#4), penetration at bottom	55 feet
West pier (#4), number of caisson anchors	24
East pier (#5), total height	247 feet
East pier (#5), depth of water	135 feet
East pier (#5), penetration at bottom	90 feet
East pier (#5), number of caisson anchors	32
Weight of each caisson anchor	570 tons
Size of caisson anchors	12 x 12 x 51 feet 6 inches in 900 foot diameter circle around caisson
Area of pier at top	118 feet 11 inches by 65 feet 11 inches
Reinforced concrete in piers	111,234 cu. yds.
Distance between piers	2,700 feet
Distance to shore from east pier (#4)	650 feet
Distance to shore from west pier (#5)	1,030 feet
Construction cost, west pier (#4)	$1,126,003
Construction cost, east pier (#5)	$1,275,723
Total pier construction cost	$2,401,726

Materials

Structural steel	10,332 tons
Reinforcing steel	1,827 tons
Cable wire	3,750 tons
Rivets	105,000
Concrete, total	103,129 cu. yds.
Concrete in anchorages	40,000 cu. yds.
Timber for pier fenders—treated	430,000 bd. ft.
Timber for pier fenders—untreated	720,000 bd. ft.

Artist's sketch of the 1940 Tacoma
Narrows Bridge, looking west.
Washington State Archives

CHAPTER FOUR

MOST BEAUTIFUL IN THE WORLD

ARCHITECTURE AND ART IN "THE MACHINE AGE"

In the 1920s, Americans embraced a robust pride in the nation's rising wealth, high standard of living, and growing technological achievements. It was a new age of airplanes, cars, nation-wide radio networks, and mass-produced (therefore more affordable) products. People took a keen interest in the arts, including architecture. There was much public debate about the design of buildings and bridges in regard to appearance and relationship to the surrounding environment.

Even after the 1929 stock market crash and the hard times of the Great Depression, the dream of greatness persisted. In 1930, poet Hart Crane published *The Bridge,* an epic work filled with dense iconography and imagery of America's past that he hoped would bridge the troubled present and link national myths and historical symbols to a wondrous, productive future.

In the 1920s and 1930s, skyscrapers, not bridges, dominated the urban environment, and skyscraper aesthetics played a significant role in the broad world of industrial and commercial design. The influential architect Louis Sullivan had designed taller and taller buildings, while emphasizing the steel skeletal frame in a skyscraper's appearance. Sullivan spoke of "the force and power of altitude," and of these immense buildings "rising in sheer exultation."

The "frozen fountain" was a metaphor used by skyscraper designers. In *The Frozen Fountain* (1932), Claude Bragdon noted, "In the skyscraper, both for structural truth and symbolic significance, there should be upward sweeping lines to dramatize the engineering fact of vertical continuity and the poetic fancy of an ascending force in resistance to gravity—

a fountain." The towers of New York's largest suspension bridges soared as high as many skyscrapers; consciously or unconsciously, suspension bridge designers were paralleling the architects' visions.[1]

In the "machine age" of the 1920s and 1930s, style had experienced significant shifts. The synthesis of beauty and use distinguished the modernist trend. The "Modern" style of the 1930s was spare, clean, direct, pure, and simple, with few embellishments or extra details, employing the subtle repetition of shapes and proportions. Designers of the new forms emphasized symmetry, simple lines, geometry, energy, texture, light, and color. New materials such as steel, chrome, sheet metal, and plastic were preferred.

The objectives of design, declared social commentator Sheldon Cheney, were "efficiency, economy, and right appearance." In *The New World Architecture* (1935) and *Art and the Machine* (1936), Cheney expressed many of the notions of the emerging "modernist" trend in the arts and structures as "aesthetic expression grounded on functional mechanized form." Industrial design was a combination of mass production, technological innovation, and abstract art. The machine age's combinations of "rational and aesthetic" elements were hallmarks of the modernist trend, said Cheney. The distinguishing marks of style were a blend of the "engineering functional rightness" of an object, and "the artist's confirmation of that rightness." The primary values were "honesty, simplicity, and functional expressiveness." A manufactured product's basic structural anatomy was "the basic design fact," because it was "beautiful in its own machine-age way." Ornamentation or decorative treatment should not conceal the nature of a radio, automobile, or refrigerator.[2]

Sketch of the 1940 bridge from the "Official Opening" pamphlet.
Washington State Archives

New York's Waldorf-Astoria Hotel, a classic Art Deco skyscraper dating from the early 1930s.

Art Deco influences became increasingly evident in popular tastes. The Deco design movement affected everything from skyscrapers and planes to furniture and appliances— and bridges. Many Deco buildings and other structures of the early 1930s displayed streamlined curves and geometric patterns. For example, New York's Empire State Building, Chrysler Building, and Waldorf-Astoria Hotel incorporated the terraced or stepped pyramid as the dominant form.

The rapid growth of the airline and automobile industries coincided with a popular fascination with speed, and streamlined airplanes and cars became the most conspicuous symbols of the new machine age. For suspension bridges, smooth flowing lines with an aerodynamic look were dramatic and economical, and gave structures the appearance of solidity and dynamic strength.

The streamlined ferry *Kalakala*, plying the waters of Puget Sound, was a floating example of Art Deco design in transportation. Built in 1935, the *Kalakala* had the honor of making the last ferry run across the Narrows on July 2, 1940, when the Tacoma Bridge opened.

More than five decades later, Sheldon Cheney's design objectives still were echoed by engineer-author David P. Billington. In "Creative Connections: Bridges as Art" in the March 1990 issue of *Civil Engineering*, Billington declared: "There are three principles of good bridge design: efficiency, economy and elegance."[3]

SUSPENSION BRIDGE ARCHITECTURE IN THE 1930S

In the late 1920s and 1930s, many long-span suspension bridges were built across the country. This reflected several important developments—first, innovations in engineering theory and practice, especially the "Deflection Theory" as applied by designer Leon Moisseiff; second, improvements in materials, particularly the introduction of high strength steels; and third, the broader society's demand for more and more bridges, spurred by the rapid rise of the automobile lifestyle.

In addition, popular taste favored more "graceful" and "elegant" bridges, which were achieved by "lightness" and "slenderness" in the new designs. At the same time, the public and the government wanted more efficient and less expensive structures in the wake of the Great Depression.

By the early 1930s, lighter and narrower suspension bridges were theoretically sound, cheaper to build, and more beautiful. Suspension bridge engineering design was summarized in the following elements—the towers deserved the "best aesthetic treatment" for the public's eye, and the suspended structure should be visually simple and structurally flexible.[4]

"MOST BEAUTIFUL IN THE WORLD"

"The most beautiful in the world," Leon Moisseiff said of his 1940 Tacoma Narrows Bridge.

The Narrows Bridge, view from the centerline. *WSDOT*

This is much more than one engineer's opinion about his own work. It reflected important architectural design trends and artistic tastes of the 1930s, as well as the culmination of three decades of suspension bridge design.[5]

Moisseiff cared deeply about aesthetics. Bridge designs, he said, needed to be "safe, convenient, economical in cost and maintenance and at the same time satisfy the sense of beauty of the average man of our time." Moisseiff believed that engineers should try "to develop the beauty of their structures" by emphasizing "the essential, to interrupt rhythmically the monotonous and to indicate the minor importance of the auxiliary…and attain the pleasure of good form." Bridge designers, he said, should "search for the graceful and elegant."[6]

The Narrows Bridge was the epitome of a modernist span. Its Art Deco and "streamlined moderne" features placed it solidly in the artistic design trends of the 1930s.

DESIGN ELEMENTS OF THE 1940 NARROWS BRIDGE

The geography of the Narrows offered a unique challenge. Here, high bluffs topped by tall evergreens straddled each side of the passage. The setting demanded an epic, even poetic, design that combined structural-technical purposes with an aesthetic impact. With its slim towers, shallow stiffening girder, and Art Deco influenced anchorages, the bridge was the essence of grace and artistry. Observers referred to the span as "her" or "she"; it was "slender," "ribbon-like," "elegant," and "graceful."

The bridge was the ultimate embodiment of Moisseiff's deflection and lateral displacement theories—aesthetically, mathematically, and economically. It surpassed all his previous designs, advancing certain patterns from the 1937 Golden Gate Bridge and the 1939 Bronx-Whitestone Bridge to a higher level of refinement, subtlety, and sophistication. Moisseiff's bold and innovative vision of the

Narrows Bridge kindled the imagination. The span would be more than a bridge, more than a crossing. His design majestically served a great artery, reaching between two evergreen-clad shores. It was not merely a crossing for cars and trucks, but an artistic and engineering statement—the culmination of his outstanding career.

Several features of the bridge place it squarely in the Art Deco tradition. Of course, all of these components had a primary practical engineering function, but the aesthetics of the design were a significant consideration.

View of a tower and the portal bracing.
WSDOT

Towers

The strong, clean appearance of the slim towers particularly characterized Moisseiff's design beliefs, dramatically reflecting his aesthetic sense and following the modernist preference for simple geometry and uncluttered lines. In the slender towers and gentle curve of the cables there was no glitz, no decorative obstruction. The visual journey was upward, enhancing the drama and splendor of the natural

The 8-foot-deep plate girder.
Washington State Archives

East anchorage sketch.
Washington State Archives

many contemporary skyscrapers. In fact, a close look at early sketches for the towers' horizontal braces reveals a virtual imitation of the pattern on the Golden Gate's towers. The vertical ribs also emphasized upward thrust, as if the towers were "frozen fountains," lifting the eye and spirit.

The corners of the horizontal braces, with stepped-off giant rectangular portals, also recalled the Golden Gate Bridge's Art Deco embellishments. The shape of the portal opening could also be viewed as being similar to a skyscraper's terraced, stepped back pyramid shape.

Solid Plate Girder

Vertical ribs on the solid plate girder likewise echoed the vertical motif. These connecting points between the steel plates stretched the full length of the 5,000-foot long girder at 8½ foot intervals.

Anchorages

Anchorages often are overlooked, but artistically they can be as important as the towers. The anchorages of the Narrows Bridge reflected the economy and simplicity seen prominently in the towers. The geometric lines of the anchorages exhibited a design that was functional and refined, without embellishment or decoration. The massive concrete form was shaped with specific

environment. The steel shafts soared skyward, with long lines and slim vertical ribs seemingly extending and amplifying the sunlight.

The towers were battered, tapering from a 50-foot base to 39 feet at the top. This also emphasized upward thrust and exaggerated the appearance of height. The tower legs offered sets of clean, vertical lines from the base to the top. It is interesting to note that Clark Eldridge and his team of designers tried to dissuade Moisseiff from using this design for the tower legs, arguing that "so many vertical lines become tiresome and detract from the whole."[7]

The vertical rib-effect on the tower legs and horizontal brace struts echoed the Golden Gate Bridge, as well as the vertical lines of

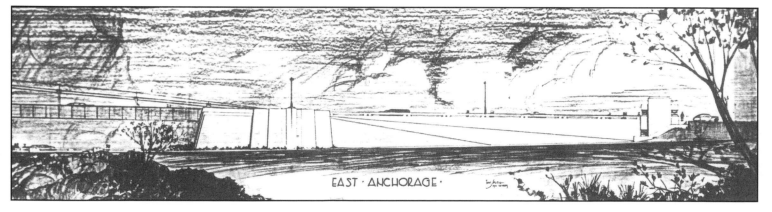

EAST · ANCHORAGE ·

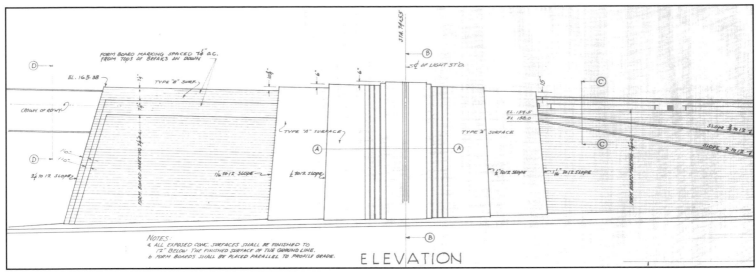

ELEVATION

East anchorage elevation.
WSDOT

·VEST·ANCHORAGE·

West anchorage sketch, 1939.
Washington State Archives

Anchorage of the George Washington Bridge,
New York, 1931.

imagery in mind. Here, the architectural treatment visually brought the cables to an ending point where they disappeared from sight. The bold lines, with power and economy, conveyed the presence of the tremendous tension of the cables. At the same time, the shape suggested a dynamic counterbalance to the cables' great pull.

The geometric motifs on the anchorages resembled two New York spans that Moisseiff worked on, the 1936 Triborough Bridge and the 1931 George Washington Bridge, both designed by Othmar Ammann. The influence of Art Deco geometry is clear when comparing the Narrows Bridge's anchorages with these bridges.[8]

Triborough Bridge anchorage, New York, 1936.

East Entrance—Toll Plaza, Light Posts, and Signage

Seen here, too, were modernist and Art Deco influences. The light posts featured three curved bands at their bases, a pattern repeated at the toll plaza in front of each toll booth. On the right side of the bridge's east entrance, motorists were greeted by the words "Tacoma Narrows Bridge," and on the left "Gateway to Olympics." The rounded lettering of these concrete signs, located near the toll plaza, also reflected styles of the period. The curved shapes created a sense of flow and movement toward the dramatic towers.

East entrance, looking toward the toll plaza.
Washington State Archives

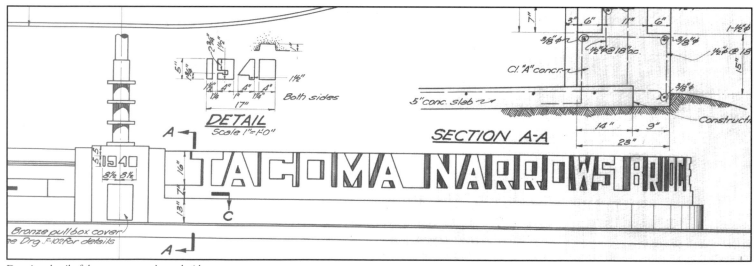

Drawing detail of the east entrance's north side.
WSDOT

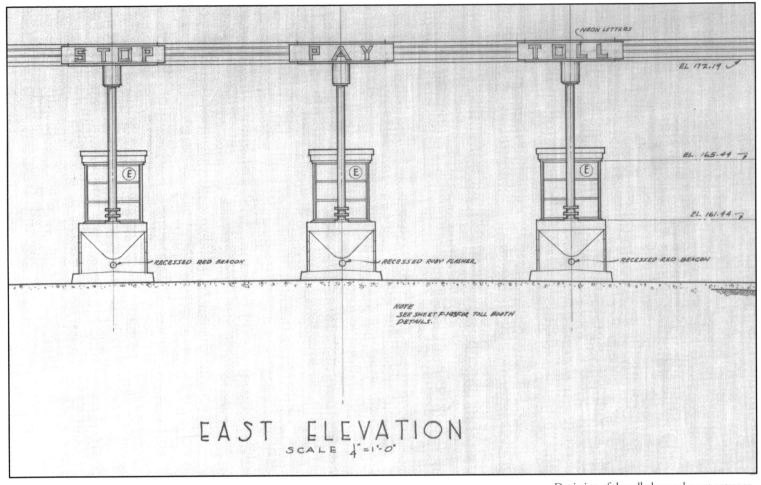

STOP   PAY   TOLL

NEON LETTERS

EL 172.19

EL. 165.44

EL. 161.44

(E) (E) (E)

RECESSED RED BEACON RECESSED RUBY FLASHER RECESSED RED BEACON

NOTE
SEE SHEET F-103 FOR TOLL BOOTH
DETAILS.

EAST ELEVATION
SCALE 1/4"=1'-0"

Depiction of the toll plaza at the east entrance.
WSDOT

Notes

1. Thomas A.P. Van Leeuwen, *The Skyward Trend of Thought: The Metaphysics of the American Skyscraper* (Cambridge, Massachusetts: MIT Press, 1988), 130–49.
2. Sheldon Cheney, *Art and the Machine* (New York: McGraw-Hill, 1936), 15–23 (quote p. 15).
3. David P. Billington, "Creative Connections: Bridges as Art," *Civil Engineering* 60 (March 1990): 50–53.
4. In the late 1930s, the dominant guiding ideas were stated by eminent suspension bridge designer Othmar Amman, in "Planning and Design of Bronx-Whitestone Bridge," *Civil Engineering* (April 1939): 217–20; David P. Billington, "History and Esthetics in Suspension Bridges," *Journal of the Structural Division, ASCE* (August 1977): 1655–72, and (March 1979): 671–87.
5. Leon Moisseiff quoted by the *Tacoma News Tribune*, November 11, 1940.
6. Leon Moisseiff, "Esthetics of Bridges," *Engineering News-Record* (November 15, 1928): 741.
7. Letter, Clark Eldridge to Leon Moisseiff, September 7, 1938, Box 42, WTBA, WSA.
8. The sketches for the east and west anchorages may have been drawn by Clark Eldridge. Letter, Clark Eldridge to Leon Moisseiff, August 2, 1938, Box 42, WTBA, WSA.

Adjusting cable wires to make a strand,
February 1940.
Washington State Archives

CHAPTER FIVE

PEOPLE OF THE 1940 NARROWS BRIDGE

THE WORKERS

Building a great suspension bridge at the Narrows took a small army of people. After the authorization and financing were finalized, engineers and draftsmen prepared detailed drawings and long lists of specifications, which were translated from paper to the physical realm at the construction site by the "dimension control" group. This team of civil engineers and surveyors measured distances with levels, transits, rods (for elevation readings), and chains (metal measuring tapes), and directed supervisors and workers for the precise placement of materials. As workmen assembled the structure, an "inspection control" group, in coordination with the dimension control group, monitored the accuracy of the construction, from measured drawings to the finished structure. Meanwhile, a large administrative staff supported all of these people.

The Narrows Bridge was completed in only 19 months—half the normal time for a suspension bridge of that size. Round-the-clock work, seven days a week, was required to remain on such a tight schedule. Crews changed shifts at 6:00 A.M., 2:00 P.M., and 10:00 P.M., and each man put in a five-day 40-hour week. The total number of workers on the job averaged around 200. During the pier construction, 190 men were employed. Some 225 men plus a couple dozen engineers and support staff worked on the bridge through early 1940. As the project raced toward completion, the average daily employment totaled 373 men.

The 1940 Narrows Bridge engineering team for the Washington Toll Bridge Authority. Left to right: Victor Haner, Harvey Donnelly, Fred Dunham, Lawrence Starkey, and Clark Eldridge.
WSDOT

Wages averaged $1.35 per hour. Steam shovel operators were near the top of the pay scale at $1.85 per hour, while laborers and apprentices, at the bottom of the scale, earned $1.00 an hour. Carpenter, cement, and pile driver foremen received around $1.60 on the average. Divers garnered the highest salary in the first year—$40 per dive and $1.33 an hour during stand-by time.

During pier construction, divers worked day and night at their hazardous jobs. A tugboat positioned and managed a diving barge at the work site. Five men in 200-pound diving suits and their "tenders" (topside assistants handling the air supply and safety lines) stood ready at all times. On an average day, a diver might make up to four trips to the bottom, working for periods of 30 to 70 minutes. Because of the Narrows' strong currents, they could only dive during slack tide. The strenuous work in 120-foot-deep water was exhausting and sometimes dangerous.

Diver Johnny Bacon, a 15-year veteran bridge worker taking a break in his 200-lb. diving suit, May 1939.
Washington State Archives

Divers checked the progress of the positioning of anchors, and of the caisson work as men on the surface poured concrete. When two of the 600-ton concrete anchors broke upon striking the bottom, divers had to wrap heavy cables around the blocks to hold them together. Underwater, visibility might be as much as 100 feet, but often it was extremely dark and the divers worked mainly by feel, not by sight.[1]

Constructing the bridge at record speed required a nucleus of experienced men. Most were ironworkers, members of the International Association of Bridge, Structural, and Ornamental Ironworkers Union. They performed a variety of jobs, including assembly, welding, and riveting of the towers, girders, deck, and other bridge elements, and they set "re-bar" (reinforcing steel) for the concrete work. Many of them traveled across the country, going from one large construction site to the next. They tended to stick with the largest bridge projects when possible, but they also worked on skyscrapers or even

hydroelectric dams if no bridges were being built, and if the work was challenging and promised overtime.

As they followed building booms all over the nation, somewhere along the line they earned the moniker "boomers." Many who came to Tacoma in 1938 had worked on the San Francisco-Oakland Bay Bridge, completed in 1936, and the Golden Gate Bridge, finished in 1937. Boomers often had learned their skills from their fathers, who in turn had learned from their fathers. Bridgemen took great pride in their work and formed the nucleus of the construction crews, lending their expertise to help guide inexperienced local workmen.

In *The Bridge*, a fascinating account of the building of the Verrazano-Narrows Bridge during the early 1960s, author Gay Talese provided an unforgettable portrait of boomers. Talese's description is just as accurate for the men who built the Narrows Bridge. The boomers on the New York span did the same kind of work and exhibited the same habits and lifestyles as their counterparts a full generation earlier.

> They drive into town in big cars, and live in furnished rooms, and drink whiskey with beer chasers, and chase women they will soon forget. They linger only a little while, only until they have built the bridge; then they are off again to another town, another bridge, linking everything but their lives.
> They possess none of the foundations of their bridges. They are part circus, part gypsy—graceful in the air, restless on the ground; it is as if the wide-open road below lacks for them the clear direction of an eight-inch beam stretching across the sky six hundred feet above the sea.[2]

Boomers were a rough lot. They worked hard, drank hard, and played hard. They often were big men with weathered complexions, proud if not cocky, and fearless, at least when it came to heights. Some had no families, but others did, usually living hundreds of miles away. Many had colorful personalities and nicknames to match. Bill Matheny, who worked on the Narrows Bridge, vividly recalls

Setting a tower base plate, August 1939. Project Engineer Clark Eldridge stands at the far right.
WSDOT

two men from the south called "Big Alabam" and "Little Alabam."[3]

One story about the bridgemen may be a myth, but it captures something of the essence of these men and the times. Long-time Gig Harbor resident Jean Robeson heard this tale as a girl. In late 1939, soon after workmen completed the first catwalks on the Narrows Bridge, one of them rode a motorcycle from one end of the unfinished bridge to the other—and lived to tell about it. However, Bill Matheny, who was there and knew many of the workers, says, "I never heard that, and I think I would have if it had happened." More likely, a boomer spun the tale over a beer at a local tavern.[4]

ONE WORKER DIED; ONE PAINTER WAS LUCKIER

Bridge construction often was dangerous and the men worked in an era of minimal government safety regulation. Hazards were everywhere and the only safety came in having good balance and common sense. Falling objects could hit workers if someone above was not careful and dropped a tool or bolt. To the public who came to watch the construction, the men looked like mere specks silhouetted against the sky. Some, especially the younger ones, were particularly daring. Workers normally had no safety harnesses, although official visitors had to take the precaution of wearing one, such as the *Tacoma News Tribune*'s Howard Clifford when he walked up one of the cables to take photographs.

The boomers remembered that 14 men perished during the Golden Gate Bridge project, and before that 23 men died building the San Francisco-Oakland Bay Bridge. Ironworker Alfred Zampa fell from the Golden Gate Bridge in 1936 during construction. He was 31 years old at the time and had "worked iron" from the time he was 20. He was caught by the bridge's massive safety net and lived, becoming a member of what was called the "Halfway to Hell Club." He not only survived, but continued to build bridges. In 2000 at age 95, Zampa attended groundbreaking ceremonies for the Carquinez Strait bridge in California's bay area that now bears his name. It is the only bridge in the United States named in honor of an ironworker or any other construction tradesman.[5]

Despite being built in record time, the Narrows project had a perfect safety record until three days before the official opening. On June 27, 1940, the first and only death during construction sent a shock through the ranks of the crews. Everyone, especially state safety inspector O.W. Eckman, had believed that the project would post a record of no fatalities. When workers had built the piers, not one man fell into the fast-moving tides. Only minor injuries occurred when constructing the towers and the suspended structure and in spinning the cables. One worker lost a finger, another broke an arm, and there were scores of bruises.

Then on June 27, as carpenter Fred Wilde stood on a 6-inch timber, he lost his balance and tumbled down a 12-foot slope, hitting his head. Co-workers rushed to his aid, but Wilde was dead. Oddly enough, the very next day a 26-year-old painter, Pete Kreller, fell 190 feet into the Narrows and survived with relatively minor injuries. Bridge engineers told Kreller that his tumble lasted four seconds and he had reached a speed of 60 mph.[6]

A BOOMER: JOHN ADOLFSON

One of the boomers was John Adolfson. By the time the bridge neared completion, Adolfson had been a boomer for some 20 years. He was 48 years old and married with one son. Before the Narrows job, he had worked on the Golden Gate and George Washington bridges.

Adolfson comfortably worked at heights that would make most people dizzy. Walking on a single cable (with hands on a second cable for balance) some 300 feet above the Narrows was a perilous part of his job. Dangerous and difficult situations were routine

John Adolfson tightens a cable strand, March 1940.
Washington State Archives

to boomers like Adolfson. He once told a reporter, "You get over thinking about falling after you've been up there a few years."

After the Narrows Bridge opened in the summer of 1940, John Adolfson and his family headed off to the next job, in Hawaii. He was building dry docks at the Pearl Harbor Naval Base when the Japanese attacked on December 7, 1941. His fate remains unknown.[7]

A Riveter: Tom "Pinetree" Colby

Tom "Pinetree" Colby was born into a Wyoming sheep-herding family in 1912. He eventually left for California, learned the riveting trade in Sacramento, and worked on the Golden Gate Bridge. He turned up at the Narrows in 1939 when the steel towers began to rise above the piers. Colby quickly earned respect for his genial personality, broad smile, and skill with hot rivets, and would later work on the 1950 Narrows Bridge.

After the workers assembled the truss-deck system, the riveting gangs followed, typically in four man crews. The job of the "heater" was to heat up and toss red-hot rivets, or send them through a pneumatic tube if the distance was too great, to a "catcher," who received and placed the rivets. Next, a "riveter" manning a rivet gun stood on one side of the steel to be joined, while on the other side stood the "bucker-up," using an air jack or hand tool to backstop the hot rivets.

Fellow worker Joe Gotchy described Pinetree Colby as a "riveter's bucker-up extraordinary." Time and again, Colby squeezed into the most uncomfortable spots at odd angles, enduring the rivet gun's loud hammering. He always came up for fresh air with a big, toothy grin. Gotchy called Colby "unforgettable" and "homely." But the rivet man's skill and unfaltering smile, even in the harshest circumstances, won the admiration and friendship of his co-workers. "Here is a diamond in the rough," Gotchy declared.[8]

Marie Guske.
WSDOT

Tom "Pinetree" Colby.
Gig Harbor Peninsula Historical Society, Photo by Joe Gotchy

The Woman: Marie Guske

The workforce typically included more than 200 men, but only one woman was employed on the Narrows project. A 20s-something blonde, Miss Marie Guske was a 1938 graduate of Washington State College. She handled secretarial work and answered telephone calls for Clark Eldridge, the lead project engineer.

CLARK ELDRIDGE (1896–1990)

Eldridge came from the small western Washington town of Lake Stevens and became one of the state's most noted engineers. A bright, energetic, and dedicated student, Eldridge completed his engineering degree in 1920 at Washington State College, and during the following 15 years distinguished himself in the Seattle City Bridge Engineer's Office. In 1936 at the age of 40, Eldridge joined the State Highway Department. Within two years he was leading the design efforts on two colossal spans, the Lake Washington Floating Bridge and the Tacoma Narrows Bridge.

Early in the Narrows project, Eldridge came to feel that it was "his bridge," as he later said. When his boss, State Highway Director Lacey V. Murrow, challenged Eldridge to find money to build it, Eldridge met with Tacoma and Pierce County officials, who agreed to turn the project over to the Washington Toll Bridge Authority and support its funding. Eldridge set to work preparing design plans and cost estimates. In the spring of 1938, Murrow took Eldridge's proposal to the Public Works Administration in Washington, D.C., where federal officials decided the plan was too expensive. Instead, they required the WTBA to hire a consultant, suspension bridge engineer Leon Moisseiff of New York, to reduce costs. The bridge was built with Moisseiff's design for the structure and Eldridge's plan for the piers, with Eldridge supervising construction.

A little over four months after completion of the bridge, Eldridge received a startling telephone call on November 7, 1940. "I was in my office about a mile away when word came that the bridge was in trouble," he later wrote. Eldridge drove to the bridge, but had to be led off the bouncing roadway by a fellow engineer. "There," he reported briefly, "we watched the final collapse."

Eldridge remained with the Washington Highway Department for only another half-year. In April 1941, he took a job with the U.S. Navy on Guam, seven months before the United States entered World War II. In December, the Japanese invaded Guam, capturing Eldridge and 71 others. He spent the rest of the war (three years and nine months) in a prisoner of war camp in Japan. On one occasion, a Japanese officer who had been a student in America recognized Eldridge. He walked up, looked at the engineer, and said simply, "Tacoma Bridge!"

After returning to the United States in 1945, Eldridge worked as a consulting engineer until his retirement in 1970. Always exhibiting great energy, Eldridge had "retired" from public employment, but continued to work until his death in 1990. Galloping Gertie remained a source of sadness in Eldridge's life. In a short unpublished memoir, he wrote, "I go over the Tacoma Bridge frequently and always with an ache in my heart. It was my bridge."[9]

Clark Eldridge.
WSDOT

LEON MOISSEIFF (1872–1943)

He was called "a man of a thousand triumphs and one failure." The lead designer of the Narrows Bridge, Leon Salomon Moisseiff, stood at the peak of his engineering profession when the ill-fated span collapsed into Puget Sound that November day.

Born in 1872 in Latvia, Moisseiff moved to New York with his parents at the age of 19, and graduated from Columbia University in 1895. Three years later, the talented young engineer joined the New York City Bridge Department, where he worked for the next 17 years. Moisseiff helped design and build some of the world's largest suspension spans, beginning with the 1909 Manhattan Bridge over the East River. He published an article about his involvement in the project that promptly won him national acclaim as the leading proponent of the Deflection Theory, which he had introduced from Europe.

Moisseiff's elaboration of the Deflection Theory laid the groundwork for the next three decades of lighter, narrower, and more

Leon Moisseiff

flexible long-span suspension bridges. These structures not only were more "graceful" and beautiful in the public eye and to engineers, they also were cheaper to build because they required far less steel than earlier spans.

Moisseiff eventually became a private consultant and was involved in the design of almost every major suspension bridge erected in the 1920s and 1930s. In recognition of his contributions, Moisseiff received various honors. The American Society of Civil Engineers awarded Moisseiff its Norman Medal and the James Laurie Prize (1939). The National Association of Manufacturers recognized him with the Modern Pioneer Award, and his alma mater, Columbia University, awarded him the Egleston Medal.

The culmination of Moisseiff's work was the 1940 Tacoma Narrows Bridge, which he called the "most beautiful" in the world. Unfortunately, Moisseiff had underestimated the importance of aerodynamics in his designs; as his bridges became lighter and narrower, they also became more flexible and unstable. When Galloping Gertie collapsed, the disaster effectively ended his career.

In 1942, Moisseiff became a member of the Advisory Board on the Investigation of Suspension Bridges. He was to chair a subcommittee of the board, but his health had been compromised since 1935 when he suffered a heart attack. On September 3, 1943, he died of heart failure at his summer home in Belmar, New Jersey, at age 71, just three years after the failure of his "most beautiful" bridge.

Tributes to Moisseiff published in the *Engineering News-Record* two weeks after his passing reflected a personal side that often is overlooked. He was a devoted family man, having married Ida Assinovsky at the age of 22, a year before he graduated from Columbia University. They had three children: a son, Siegfried, and two daughters, Liberty and Grace. An educated and cultivated man, Moisseiff loved art, music, and literature. Colleagues and friends praised Moisseiff for his "keen analytical mind, professional honesty

and moral integrity." He was modest, avoiding the limelight and worked quietly "without fanfare."

Moisseiff impressed others with his "keen sense of humor." He was by nature "broad-minded and charitable," considerate, and unselfish. In recognition of his contributions to the engineering profession, the American Society of Civil Engineers established the annual Moisseiff Award for the most outstanding paper published in the field of structural design.[10]

LACEY V. MURROW (1904–1966)

As the younger brother of Edward R. Murrow, the noted radio and television journalist, Lacey V. Murrow never achieved the same level of public recognition as his older brother. Still, he made important contributions to history and received numerous honors during his lifetime.

Murrow was born in Greensboro, North Carolina, in 1904. The family soon moved to the Pacific Northwest, near Bellingham. After graduating from high school, Murrow worked intermittently for the State Highway Department. Like his two brothers, Lacey attended Washington State College. In 1923 he graduated with a bachelor's degree in civil engineering. He enjoyed flying and soon earned a pilot's license as well.

Murrow returned to full-time employment at the State Highway Department and quickly moved up to positions of higher responsibility. In 1933, at the age of only 29, Murrow began an eight-year term as director of the State Highway Department. It proved to be a turning point, both for his career and for bridge building in Washington. When the Washington Toll Bridge Authority was created in 1937, Murrow also served as its chief engineer. He successfully advocated a floating bridge across Lake Washington, which was completed in 1940, and simultaneously guided completion of the Tacoma Narrows Bridge.

Lacey V. Murrow.
WSDOT

World War II took Murrow's career in a new direction. His pre-war flying hobby now became his duty in wartime. As a military pilot from 1941 to 1946, he won several honors, including a Presidential Citation with four cluster decorations, the Legion of Merit, the Order of the British Empire, and the Croix de Guerre. In 1951, Murrow was promoted to Brigadier General in the U.S. Air Force. He served in Korea, Japan, and the United States before retiring from the military in 1954. He remained in Washington, D.C., where he worked for Transportation Consultants Inc., serving most of those years as the firm's president, until his death.

Lacey Murrow, like his brother Edward, smoked cigarettes from the time he was a young man. In the early 1960s, also like Edward, he began suffering from lung cancer. In 1965, Lacey underwent an operation shortly before his brother's death from the disease. Just over a year later, Lacey was found dead at the age of 62 in his room at the Lord Baltimore Hotel. A 12-gauge shotgun stood against the bed.

Shortly after Murrow's suicide, the Washington State Legislature requested that the State Highway Commission rename the first Lake Washington Floating Bridge in his honor. In March 1967, the commission paid tribute to Murrow, declaring "this notable engineering achievement [had] received world-wide recognition for its pioneering of a new concept in over-water structures."[11]

ALFRED G. SIMMER (1878–1953)

Known as "Simmie" to his friends, photographer Alfred Simmer left an extraordinary legacy. His artistry with the camera provided excellent photographic documentation of the Narrows Bridge project, as well as many other of the state's historic spans.

Born in Danzig, Germany, he attended school in Budapest, Hungary, during his early years before moving to San Francisco with his parents at age 13. He always was an insatiably curious person, eager to expand his knowledge of his surroundings. When he set about learning the language of his new home, he became an English teacher in little more than three years. Eventually, he took a teaching job in Port Angeles in 1894. In his spare time, Simmer studied electricity, telephone operations, then civil engineering, drafting, and architecture.

A latent talent at artistic lettering emerged, which he turned into a commercial venture, producing a little part-time income. In 1903, he married Austrian-born Mary Louise Setril in Seattle, and the couple moved to Nome, Alaska, where Simmer worked for a telephone company, installing the territory's first switchboard. He also built houses, including one that he and his wife occupied.

Having gained a wealth of construction and industrial knowledge and skills, Simmer returned to Seattle in 1910. He now had a new hobby, photography, which soon become his profession. He began doing photographic work for construction and architectural firms. In 1912, he moved his wife, son, and daughter to Wenatchee, where he operated a photography studio for 17 years. In 1939, Simmer returned to western Washington to work for the State Highway Department. He spent the next 13 years with the Bridge Division, then the Planning Division, leaving a valuable pictorial record. Simmer retired in 1952 and moved to California to be near his children (his wife had passed away in 1946). A few months later, he died at the age of 75 in January 1953.

Alfred Simmer's colleagues at the State Highway Department regarded him highly, as much for his personal traits as for his excellence as a photographer. A friend later described him "as a rugged individual…yet always tender, considerate, polite…His kindly, sparkling smile, his quick youthful step, his effervescent good humor and his ability to observe the little problems of the other fellow and to ease his burden as much as he could have notched a pathway deep into the memory of many who recognize in our Simmie the fighting heart of a great man."[12]

Notes

1. "Spinning a Web of Steel High above Puget Sound," *Seattle Times*, February 18, 1940; "Divers Worked in 120 Feet of Water," *Tacoma News Tribune*, July 1, 1940; Joe Gotchy, *Bridging the Narrows* (Gig Harbor, Washington: Peninsula Historical Society, 1990), 17–38; "Tacoma Narrows Bridge, Tacoma Washington: Report on Construction of the Substructure," by H.F. Connelly, 1940, Box 18, WTBA, WSA; "Tacoma Narrows Bridge, Tacoma Washington: Report on Construction of the Superstructure," by H.F. Connelly, 1940, Box 18, WTBA, WSA.

2. Gay Talese, *The Bridge* (New York: Walker, 1964), 1. Another excellent portrait of bridge workers is Peter Stackpole's *The Bridge Builders: Photographs and Documents of the Raising of the San Francisco Bay Bridge, 1934–1936* (Corte Madera, California: Pomegranate Artbooks, 1984).

3. Bill Matheny, interview, 2005; Talese, *The Bridge*, 1–11.

4. Jean Robeson, interview, 2003; Bill Matheny, interview, 2005.

5. John V. Robinson, *Al Zampa and the Bay Area Bridges (Images of America)* (San Francisco: Arcadia, 2005).

6. Lacey V. Murrow, "Construction Starts on the Narrows Bridge," *Pacific Builder and Engineer* 45 (March 4, 1939): 34–35; "Lone Death Shock to All Bridge Folk," *Tacoma News Tribune*, July 1, 1940.

7. "Spinning a Web of Steel High above Puget Sound," *Seattle Times*, February 18, 1940.

8. Gotchy, *Bridging the Narrows*, 77–78; "Spinning a Web of Steel High above Puget Sound," *Seattle Times*, February 18, 1940.

9. Clark H. Eldridge, "An Autobiography"; "Capture and Imprisonment by Japs," *Pacific Builder and Engineer* 51 (December 1945): 44–49; Dorris Hensel, "Longtime Bridge Engineer Recalls 'Galloping Gertie' with Heartache," *Daily Olympian*, September 3, 1986; "Engineer Has Long Career as Bridge Builder," *Seattle Times*, June 1, 1940; "Clark Eldridge, 94, Bridge Builder, Didn't Let Age Dull Desire to Work," *Seattle Times*, November 9, 1990.

10. "A Great Engineer," [editorial] *Engineering News-Record* 128 (September 9, 1943): 78; "Leon S. Moisseiff Dies; Famous Bridge Engineer," *Engineering News-Record* 128 (September 9, 1943): 70; Leon Moisseiff, "Growth in Suspension Bridge Knowledge," *Engineering News-Record* 123 (August 17, 1939): 46–49; Leon Moisseiff, "Esthetics of Bridges," [review of book by Friedrich Hartmann], *Engineering News-Record* 123 (November 15, 1928): 741; "Memoir of Leon Solomon Moisseiff" [prepared by O.H. Ammann and Fredrick Lienhard], *Transactions of the American Society of Engineers* 111 (1946): 1509–12; Obituary, *New York Times*, September 4, 1943; Henry Petroski, "Leon Solomon Moisseiff," *American National Biography*, Vol. 15 (New York: Oxford University Press, 1999), 662–64; Petroski, *Engineers of Dreams,* 270–72, 289–308; "Leon Solomon Moisseiff," *Dictionary of American Biography*, Supplement Three, 1941–1945 (New York: Charles Scribner's Sons), 530–31; "Tributes to L.S. Moisseiff," *Engineering News-Record* 128 (September 23, 1943): 74–75; Bart Ripp, "A Dream that Danced and Died," *Tacoma News Tribune*, October 23, 1995; B.W. Brintnall, "Strong New Span, Engineer's Promise," *Tacoma News Tribune*, November 15, 1940; "Consulting Engineers" folders, Box 42 and "Moisseiff" folder, Box 53, WTBA, WSA.

11. "Lacey Murrow, Former Director, Honored at WSC," *Highway News* 8 (May-June 1959): 19; "Ex-State Highway Director Murrow Dies of Gunshot," undated clipping, Lacey V. Murrow biographical file, Washington State Library; *Who's Who in the State of Washington, 1939–1940* (Seattle, 1940), 137–38; *America's Young Men: The Official Who's Who among the Young Men of the Nation* (Los Angeles: Richard Blank, 1934), 439; "Dean of Europe's Radio War Broadcasters Is Brother of Lacey Murrow," *Pacific Builder and Engineer* 45 (November 4, 1939): 31.

12 Bill Miller, "Simmie," *Highway News* 2 (February 1953): 3–4.

CHAPTER SIX

WARNING SIGNS AND RUMORS

FOREWARNING

Clark Eldridge and the state's engineers expected instability in the form of vertical oscillations long before construction began. In fact, Eldridge and his team were concerned about Moisseiff's design from the first time they saw it in late July 1938.

The plans also raised doubts elsewhere. The Reconstruction Finance Corporation passed Moisseiff's superstructure design to Theodore Condron, a Chicago-based consulting engineer who advised the RFC on projects that were requesting construction loans. Condron became troubled as he reviewed Moisseiff's work between July and September 1938. When Condron compared it to other recently completed major suspension spans, the proposed Tacoma bridge would be the narrowest ever built. Its width-to-center (in layman's terms, width-to-length) ratio of 1 to 72 (39 feet to 2,800 feet of the center span) was dramatically slimmer than the world's narrowest suspension structure, the Golden Gate Bridge, at 1 to 47.[1]

Condron wanted more information. In late August 1938, he visited the Pacific Northwest to confer with the engineers. The discussions left Condron feeling uneasy. Eldridge next advised Moisseiff, explaining that Condron had "also expressed concern over the extreme narrowness of the structure and the use of the plate girders instead of stiffening trusses."

The plan for the tower piers also was questioned. One of the Washington Toll Bridge Authority's consulting engineers, Ray McMinn, reported to Highway Department head Lacey Murrow, saying "the fact that the pier design turned out as it did made him [Condron] more concerned about the superstructure." Eldridge's original design for the piers had not needed changing, yet the

East end of the bridge, looking west from the Tacoma side. Note the tie-down cables attached to the side span in the foreground.
Washington State Archives

consultants, Moran & Proctor, had revised the plans. (Later, when the construction bidding opened, contractors objected to the Moran & Proctor design and Eldridge's original pier proposal was reinstated). Condron's concern about the superstructure, McMinn added, was "perfectly proper, for the bridge is without precedent in respect to lateral stiffness and height of stiffening trusses."[2]

Next, Condron traveled to Berkeley, California, to confer with Dr. R.E. Davis at the University of California, who had studied the Golden Gate Bridge. Davis reviewed Moisseiff's Tacoma Narrows design and told Condron that he felt "reasonably confident" that the span's lateral deflection would be within safe limits. Condron, however, still felt uneasy about the bridge's exceptional flexibility and decided to address the matter directly with Moisseiff.

In response, Moisseiff wrote back to Condron, stating that his design was indeed slender but the stiffening would be "rather satisfactory." Moisseiff's lukewarm endorsement of his own work failed to reassure Condron.

With a deadline for his report looming, Condron reluctantly decided to approve the design. He deferred to Moisseiff's plan, emphasizing, "In view of Mr. Moisseiff's recognized ability and reputation...I feel we may rely upon his own determination of stresses and deflections."

But Condron qualified his endorsement, saying the "unusual narrowness" of Moisseiff's plan remained a concern. Using the Golden Gate Bridge as his benchmark, Condron expressed doubts about going too far beyond the limits of contemporary engineering experience. "I recommend that serious consideration be given to the possible increase in the width of this structure, before the contract is let or work begun." Condron suggested increasing the width to 52 feet. This would increase the width-to-center span ratio to 1 to 53, creating a structure less extreme in its narrowness, less flexible, and nearer to the Golden Gate Bridge's proportions. Federal authorities, however, ignored Condron's warning and approved the Washington Toll Bridge Authority's loan request.[3]

Meanwhile in Olympia, the Board of Consulting Engineers for the WTBA, led by Charles Andrew, accepted Moisseiff's redesign specifications, although with qualifications. "We believe that the present span could be materially increased if it were necessary, keeping the same width without any detrimental effect. In consequence we have no concern as to the general features of the proposed design of the superstructure." Lacking sufficient time to examine the plans in detail and to check the stress calculations, the board pronounced its "full confidence" in Moisseiff, considering his prominence and reputation.[4]

TROUBLED BRIDGE OVER SWIFT WATERS

Consequently, before construction began Clark Eldridge and his team believed they had a troubled bridge on their hands. Meanwhile, another Moisseiff-designed plate girder suspended structure, the Bronx-Whitestone

Senator and Mrs. Homer T. Bone on a catwalk with H. Woodworth, D. McEachern, and Clark Eldridge shortly after the start of cable spinning, January 23, 1940.
WSDOT

Bridge, began to exhibit significant vertical movement during the latter stages of construction before its completion in April 1939. In February, Eldridge contacted an associate professor of engineering at the University of Washington, F. Burt Farquharson, and soon a contract for a study was arranged.

Over the ensuing months, Farquharson and Eldridge noted the behavior of other suspension bridges with relatively shallow stiffening girders. Rumors also were circulating among engineers concerning vertical waves in several recently completed spans. Most notable in this regard were the Golden Gate, Thousand Islands, Deer Isle, and Bronx-Whitestone bridges, all completed between 1937 and 1939. An alarming event occurred at the Golden Gate on February 9, 1938. During a high wind storm, the span began experiencing vertical waves "of considerable magnitude."[5]

When the Bronx-Whitestone Bridge opened in April 1939, Eldridge asked Moisseiff whether his Narrows design should be changed. Moisseiff was polite but blunt. He replied (to Murrow, not Eldridge directly) that "reports of undulation were probably erroneous and that in any event this type of motion could be easily corrected and that no change in plan was necessary." Moisseiff may have had other reasons to minimize concerns. At the time, the vertical movements of the four bridges were not widely reported because of possible negative impacts on insurance coverage and the public's confidence.[6]

In this period another warning reportedly came from Dexter R. Smith, an Oregon engineer. Smith worked for the Oregon Department of Highways and also taught civil engineering at Oregon State College in Corvallis. Moisseiff's design was sent to Smith, and he was asked for his opinion. It is unknown who sent the plans. Possibly it was Clark Eldridge, or perhaps Charles Andrew, who had formerly worked in Oregon and who might have known Smith, or maybe Ray McMinn, the WTBA's consulting engineer, who worked

for the U.S. Bureau of Public Roads in Portland, Oregon, and who also may have been acquainted with Smith. After Smith calculated the stresses, using a different method than Moisseiff's Deflection Theory, he came to the conclusion that the bridge would fail. He shared his findings with the Washington Toll Bridge Authority, but they took no action.

Driven by concerns for public safety and the reputation of the engineering profession, Dexter Smith took a bold step. At a national conference (possibly at an American Society of Civil Engineers meeting in New York) with Moisseiff and other major suspension bridge engineers in attendance, Smith publicly predicted the failure of the Narrows Bridge. But this challenge to one of America's greatest designers by an upstart engineer in his thirties seemed ludicrous. Supposedly, they shouted angrily at Smith and drove him from the meeting, and the State of Oregon later revoked his engineering license for "unethical behavior," but events would soon vindicate Smith beyond his expectations.[7]

Beginning in February 1939, the WTBA and the PWA jointly funded Professor Farquharson's studies. In the basement of Guggenheim Hall at the University of Washington, Farquharson built a model of the bridge as a tool to help develop procedures for

Professor F. Burt Farquharson, 1940.
Tacoma Public Library 9492

Full-scale bridge model in the wind tunnel at the University of Washington, September 1940.
Washington State Archives

constructing the span. He assembled a 1:200 scale model (54 feet long) of the entire bridge with the help of students. They also made a 1:20 scale model (8 feet long) of a section of the deck. He began tests on the large model, using 100 special electromagnets placed along the model's roadway to simulate wind and other forces. As Farquharson gathered information on other suspension bridges with stability problems, he decided to construct a more elaborate full model.

CORRECTING THE BOUNCE, TOO LATE

By the first week of May 1940, as workmen finished the bridge's floor system, they noticed the deck's vertical wave motions, or "bounce," and knew something was wrong. Eldridge grew increasingly concerned as the oscillations grew progressively larger. The structural instability was apparent and a remedy was needed. He began keeping detailed notes on the oscillations and faithfully maintained this log until the collapse on November 7, 1940.

Eldridge sent a telegram to Moisseiff, who soon boarded a train for Tacoma to see for himself. Moisseiff recommended the same remedies to reduce the waves that had been applied with some success by other engineers on the Deer Isle and Bronx-Whitestone bridges. He returned to New York and, at Eldridge's request, designed center ties and hydraulic jacks, to be added for increased stability. In May 1940, four hydraulic jacks, or "buffers," were installed at the towers to act as shock absorbers. But the design proved faulty and the devices operated for only a few days. In late June, when workmen sandblasted nearby metal parts prior to painting, they inadvertently destroyed the leather piston covers and the buffer devices immediately failed.

Workers also installed center ties in early June—diagonal stays that connected the midspan cable bands and the stiffening girders. Each stay was a 1½-inch-diameter steel cable with a strength capacity of 270,000 pounds.

During the first weeks after the bridge opened in July 1940, the Reconstruction Finance Corporation's inspecting engineer, James Roper, reported to Morton McCartney, Chief of the Self-Liquidating Section at the RFC in Washington, D.C. In regard to the unusual vibrations, he noted, "the total up and down movement was about 8", although I understand it has been considerably more," and the "breathing as it is called, was noticeable."[8]

Farquharson later recounted, "We knew from the night of the day the bridge opened that something was wrong. On that night the bridge began to gallop." Actually, he knew about it before then. By May 1940, it was obvious to Farquharson that the bridge was "a very lively structure," as he carefully monitored the movements, recording wind speeds and the size and shape of Galloping Gertie's vertical oscillations.

Starting in late July, Farquharson set up a motion picture camera, along with a transit and 10 black and yellow targets mounted on lamp posts at the south side of the eastern half of the bridge; six targets were located on the main span and four on the side span. He began to collect data that he hoped would solve the puzzle of Gertie's instability. From August 1 to October 10, 1940, he kept a regular log of observation dates and times, wind velocity and direction, and structural motion.

As Farquharson's tests at the university progressed and he gathered more on-site data, he grew increasingly concerned. The gentle breezes of summer soon gave way to autumn's stronger winds and he noted that the bridge moved with "ever-increasing frequency." Farquharson's conclusion that it displayed "potentially dangerous" movements had become "inescapable." By September, Farquharson completed the second full model, this one 96 inches long and 24½ inches wide. He began tests in a small 8-by-12-foot wind tunnel in the Guggenheim Hall basement at the University of Washington. He was determined to find a solution that would stabilize the bridge.[9]

Bridge engineer David Steinman, who had achieved some success in quieting similar (but less dramatic) vertical motions on the Thousand Islands and Deer Isle bridges, wrote to the Washington Highway Department on August 2 and offered his assistance. Murrow referred him to Leon Moisseiff, who declined Steinman's help as unnecessary. Later, after Galloping Gertie failed, Steinman published a bragging declaration—"Even so, the mid-span stays at Tacoma, copied from me, although inadequate were the only thing that kept the bridge from going into destructive oscillations during the four months of its life…If they had let me help them, I could have saved the bridge, as I have saved other bridges. I could have made the Tacoma span safe for a very small expenditure. But my volunteer offer went begging."[10]

That autumn several storms lashed the bridge with winds exceeding 50 mph. The bridge remained steady, however, showing no hint of what was to come. In October, while Farquharson's studies continued, engineers placed "temporary" tie-down cables (anchored restraining wires 1⁹⁄₁₆ inches in diameter) on the side spans some 300 feet out from each anchorage. The adding of tie-down wires from the tower tops to the decks was considered but never tried. Engineers believed they had solved the problem. In October, Eldridge reported to Theodore Condron about the superstructure's movements, using the term "deck flutter" and describing the corrective measures being taken. On November 1, an east tie-down cable broke in a high wind when Gertie began to "gallop" and workmen immediately began repairing it. The tie-down cables did reduce the bounce in the side spans, but had no effect on the critical center span.[11]

Farquharson's studies for the WTBA took longer than he expected, as other projects relating to federal defense research took priority. The Tacoma Narrows studies were completed November 2, although he did prepare a preliminary report of his aerodynamic analysis on October 14. The probable cause of the instability, he said, was the solid stiffening girders. The report opened on a somber note: "The result of several months of study on the dynamic model of the Tacoma Narrows Suspension Bridge seems to indicate conclusively that the best that can be expected as a result of further stiffening of the bridge will be a moderate reduction of the amplitude of the various modes of motion developed under varying conditions together with a slight modification of the frequency of each of these respective modes."[12]

After detailing the nature of his observations and the laboratory investigations, Farquharson noted, "It was clear at once that the phenomena bore a striking resemblance to the galloping transmission line described by Den Hertog," an engineer who in 1932 reported on wind-induced vibrations in electrical power lines. Farquharson offered a choice of remedies—either allow the wind to pass through by cutting holes in the solid girders, or deflect the wind by covering the girders with sections of curved steel, or "fairings." He added a chilling postscript, "even as [this] memo is being concluded a report comes in of violent action in a 49 m.p.h. wind."

Steel tie-down cables attached to anchor blocks.
Washington State Archives

Professor Farquharson concluded his studies on Saturday, November 2, 1940, perfecting the deflector device that should dampen Gertie's gallop. Near the end of his research Farquharson noted that under certain conditions there was a "twisting motion" on the model. "We watched it," he later told reporters, "and we said that if that sort of motion ever occurred on the real bridge, it would be the end of the bridge."[13]

No One Expected a Collapse... Almost No One

The Narrows Bridge had been designed by one of the world's most respected bridge engineers, and federal and state experts had approved the plans. It was a state-of-the-art structure costing more than $6 million, erected in a period when Americans placed great faith in technology. Many people also were aware that no large suspension bridge had failed for decades.

The state's engineers announced in local newspapers that the "bounce" was normal and they were in the process of installing motion damping devices and safety measures. There was no reason for the public to become alarmed by Gertie's gallop. They knew the bridge had trouble with its "bounce," but Professor Farquharson had been contracted to devise remedies. The Toll Bridge Authority expressed optimism and their delight with the revenues generated by the popular new route. The WTBA also was taking a close look at the insurance policies, hoping to replace them with ones carrying lower premiums.

Some workers were far less optimistic. As the bridge neared completion in May 1940, the ripples alarmed many workmen. Some chewed on lemons to counteract motion sickness and some believed that Galloping Gertie would go down in a matter of months. One of them, Lefty Underkoffler, who earlier had worked on the Golden Gate Bridge, said, "it swayed so much men actually got seasick working on it. When we saw the caissons

being floated in and realized how small they were, we made bets how soon the bridge would collapse."[14]

Likewise, F.S. Heffernan said to Ted Coos, "I'll bet you the bridge won't last a year." Heffernan's company, Glacier Sand and Gravel, was a supplier for the project, while Coos worked as a design engineer for the Pacific Bridge Company. Later when the bridge failed, Heffernan was as upset as everyone else. "I couldn't take the money now," he said sadly.

Early that autumn a bus loaded with civil engineering students from Oregon State College drove north to the Tacoma Narrows on a tour. One of them was Lewis Melson. "On the return bus ride back to the college," recalled Melson, "we decided the bridge would collapse within five years."[15]

November 6, the Day Before

For the average Tacoma resident, Wednesday, November 6, 1940, came and went much as usual. Trains pulled into Union Station full of soldiers headed for new assignments at Fort Lewis and McCord Air Base. It was the day after the Presidential election and newspapers reported signs of a Democratic landslide from the first scattered poll returns. On the international front, there was news of fighting in Greece following Mussolini's invasion weeks before. At local stores in Tacoma shoppers could buy a two-pound tin of coffee for 39¢ and a loaf of bread, on sale, for 13¢. That evening some people with spare change went to the movies at the Roxy, where they enjoyed *Public Debutante No. 1,* starring George Murphy and Brenda Joyce.

At the *Tacoma News Tribune* offices, 28-year-old Howard Clifford worked on a variety of assignments. Most days he filled in for others on the news staff to write a story about a car wreck or fire, or to take a few photos as a backup cameraman. The *Tribune*'s news editor, Leonard Coatsworth, carefully corrected and finalized articles for the next morning's

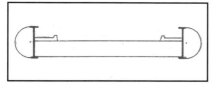

Drawing of a curved wind deflector from Professor Farquharson's October 14, 1940, report. In the wind tunnel tests, deflectors made the bridge model "very positively stable" in a simulated 50 mph wind.
Washington State Archives

edition. He planned to take the next day off to drive to his family's summer place in Arletta, just a few miles across the Narrows on the Peninsula.

Just four days after Farquharson had finished his studies, state authorities had drafted a contract to install the wind deflectors, which would cost an estimated $80,000. Eldridge had met with Farquharson and PWA engineer L.R. Durkee to decide on a course of action. They agreed on a plan for "streamlining" the south side of the center span. With approval from the Toll Bridge Authority's chief engineer, Charles Andrew, Eldridge began preparing engineering sketches and getting prices for steel and other materials. In ten days the bridge would have enough wind deflectors to achieve significant stability, if the gusts came from the south. In two weeks, the south side

of the center span would be fully covered with deflectors, and in 45 days the entire bridge on both sides would be protected. Eldridge and Farquharson felt optimistic.

That evening Carol Peacock, a Fife High School student, sat down to do her homework for a journalism class. The assignment was to let her imagination run wild and write an essay that began, "Just suppose…" She jotted down the title, "Tacoma Narrows Bridge Collapses," and completed the essay that she would turn in tomorrow at school.

The weather forecast predicted a storm overnight, with gale force southerly winds. In the early hours of November 7, the winds blew in over Puget Sound, gusting up the Narrows into the side of Gertie's eight-foot solid plate girders. Time had run out.

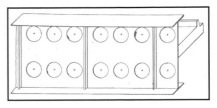

An optional remedy for Galloping Gertie's instability from Farquharson's October report—six holes, 22½ inches in diameter, cut into each 8-foot girder section.
Washington State Archives

Notes

1. Petroski, *Engineers of Dreams*, 300.
2. Letter, L.V. Murrow (by Clark Eldridge) to Leon Moisseiff, September 8, 1938, and letter, R.B. McMinn to L.V. Murrow, September 2, 1938, Box 42, WTBA, WSA.
3. Petroski, *Engineers of Dreams*, 297–300. Excerpt from Theodore Condron's report of September 21, 1938 is in Advisory Board, *Failure of the Tacoma Narrows Bridge*, Appendix IV.
4. "Report of Board of Consulting Engineers, The Tacoma Narrows Bridge," August 31, 1938, Box 42, WTBA, WSA.
5. Farquharson, *Aerodynamic Stability*, Acknowledgements and 14, 19, 41.
6. "General Report on the Design of the Tacoma Narrows Bridge," Charles Andrew, January 15, 1942, Box 53, WTBA, WSA.
7. Charles "Chuck" Munson, telephone interviews, November 2003 and January 2004; James Howland, telephone interview, February 2004. Munson and Howland are not acquaintances and contributed their accounts of this incident independently of one another. No documentation exists for the 1939 and 1940 ASCE meetings, except for papers published in the "Proceedings and Transactions." There is no record of Dexter R. Smith's reported contact with the Washington Toll Bridge Authority in the WTBA records at the Washington State Archives.
8. Eldridge, "An Autobiography," 41–42; Farquharson, *Aerodynamic Stability*, 19; letter, James Roper to Morton McCartney, July 18, 1940, copy in Box 102, von Karman Papers, Caltech; letter, L.R. Durkee to Clark Eldridge, January 11, 1941, Box 20, WTBA, WSA. Eldridge's notes on the bridge's oscillations were furnished to the Carmody Board for its investigation after the November 7, 1940, bridge failure.
9. Farquharson, *Aerodynamic Stability*, 14, 19, 41, 55–57; F. Burt Farquharson, "A Dynamic Model of Tacoma Narrows Bridge," *Civil Engineering* 10 (July 1940): 445–47; N.A. Bowers, "Model Tests Showed Aerodynamic Instability of Tacoma

Narrows Bridge," *Engineering News-Record* 125 (November 21, 1940): 674–77; "Bouncing Span Will Soon Be Quieted Down," *Tacoma News Tribune*, July 1, 1940; "Cables Will Curb Bridge's Bouncing," *Tacoma News Tribune*, September 5, 1940; "Would Bounce Bounce," *Tacoma Ledger*, October 6, 1940; "Present Project on Narrows Span to Stop Swaying only Temporary," *Tacoma Times*, October 6, 1940; "Engineers to Study 'Ripple' on Big Span," *Seattle Post-Intelligencer*, July 10, 1940; "Model of Huge Span Aids Engineers at University," *Seattle Times*, April 21, 1940; "Various Types of Motion Observed on Tacoma Narrows Bridge," January 2, 1941, report attached to letter from F. Burt Farquharson to James A. Davis, January 13, 1941, Box 60, WTBA, WSA.
10. Letter, L.V. Murrow to David Steinman, August 16, 1940, Box 42, WTBA, WSA; David Steinman letter to editor, *Engineering News-Record* 127 (December 18, 1941): 51; Steinman quoted in Scott, *In the Wake of Tacoma*, 60.
11. Letter, Clark Eldridge to Theodore Condron, November 16, 1940, Box 102, von Karman Papers, Caltech.
12. "The Tacoma Narrows Bridge Aerodynamic Analysis," October 14, 1940, Box 16, Engineering Experiment Station records, Special Collections, UW Libraries.
13. *Ibid.*; "Would This Have Saved Bridge?" *Seattle Times*, November 8, 1940; "Device Found but too Late," *Seattle Post-Intelligencer*, November 10, 1940; "Plan to Save Span too Late; Professor 'Rides' Break-up," *Seattle Times*, November 8, 1940. F. Burt Farquharson quoted in *Tacoma News Tribune*, November 8, 1940.
14. Quoted in John Van der Zee, *The Gate: The True Story of the Design and Construction of the Golden Gate Bridge* (New York: Simon and Schuster, 1986), 317.
15. Telephone interview, Lewis B. Melson, February 2004; also, Melson's letter to the editor, *Oregon Stater* (Oregon State University), February 1997.

Shortly after 10 A.M., the torsional oscillation begins. Leonard Coatsworth's car is barely visible at right-center in this view from the east shoreline.

Gig Harbor Peninsula Historical Society, Bashford 2784

CHAPTER SEVEN

GALLOPING GERTIE COLLAPSES
NOVEMBER 7, 1940

THE FATEFUL DAY UNFOLDS

In the early hours of Thursday, November 7, 1940, strong winds raced through the Narrows from the south. Around 3:30 a.m., the temporary hold-down cables were vibrating noisily, awakening a Washington Toll Bridge Authority employee sleeping on a scow beneath the west end of the bridge. Around 5:00 a.m., the wind woke another WTBA employee, Kenneth Arkin, at his house. He rose and drove to the Narrows to check the temporary tie-down cables on the side spans. He noted that the brisk wind was blowing from the south with little effect on the bridge. He drove home, ate breakfast, and prepared to return to work.[1]

7:30 A.M. Arkin arrived back at the bridge, walked to mid span, and read the wind velocity—38 mph—on an anemometer. The bridge was bouncing with several waves two to four feet high.

At his home in Seattle, Professor Burt Farquharson had risen early. He noticed the strong wind and guessed that Gertie might be bouncing. He decided to drive to the Narrows to make observations and recordings.

Around 8:30 A.M. Engineer Clark Eldridge drove across the bridge. The center span was doing its familiar wave, but less than it had on other days. He returned to his office a mile away, and by 9 o'clock began to work on sketches and the contract specifications to streamline the south side of the bridge.

About 9:40 A.M. Engineers clocked the wind at 42 mph near the bridge's east end. Close to the west end, fishermen noted the wind and later said they thought it was "substantially" higher. Professor Farquharson

arrived after an hour drive from Seattle. He donned his tan trench coat, grabbed his pipe, and began taking photographs and motion pictures of the rippling roadway for his engineering studies.

Also about this time a student from the College of Puget Sound, Winfield Brown, walked onto the moving bridge to get "a thrill for a dime." Brown reached the west tower and returned. Then he turned and again walked west, hoping to get a look at the Coast Guard vessel *Atlanta* that soon would pass beneath the bridge.

Between 9:50 and 10:00 A.M. The last cars to safely cross Galloping Gertie paid their tolls and drove westward toward Gig Harbor. The toll booth agent warned one truck driver, Jack Penny, that the bridge was "a little shaky." At this time, Elbert "Keith" Swinney was driving his regular route for the Golden Rule Bakery, crossing the Narrows from Tacoma toward Gig Harbor. In the truck were his wife, Hazel, and their five-year-old son, Richard. Just behind them came a car driven by Dr. Jesse W. Read.

Meanwhile, heading from the west side over the Narrows was postal carrier Francis Morris, who daily transported mail from the Gig Harbor post office to Tacoma's main post office. When approaching the bridge in her 1930 Model A Ford, Morris saw the bouncing, as she had on many occasions. "Thinking it a little exciting," she later recalled, "I drove on." Soon, the trip became more exciting than she could have guessed. Morris' Model A would be the last car to cross from west to east.

Those who followed were not so lucky. Just after 10 o'clock, a loaded logging truck started onto the west approach span. At the

The torsional motion continued with little change for about a half-hour.
Washington State Archives

a few minutes, the violent motion broke loose huge chunks of concrete from the curbs along the roadway, tossing them randomly across the center of the span.[2]

On the west half of the bridge, the Swinney family heading west in the Golden Rule Bakery truck careened along the road as conditions worsened. It was an exciting trip for five-year-old Richard. Years later he recalled, "The sides of the bridge were solid walls. I would see water on one side and then the other. My mother was screaming a lot." They made it safely across and only later learned about the collapse. Just behind them, Dr. Read was the last person to safely cross from east to west. He stopped at the west end, got out of his car, and began taking movies with his 8-mm camera.

Near the west tower sat the Rapid Transfer Company van with passengers Ruby Jacox and Walter Hagen. Unable to drive forward to the west end and trembling with fear, they jumped from the van just seconds before the tilting roadway tipped the vehicle over. The two clung to the curb. Closer to the west tower, the logging truck pitched back and forth, then suddenly its load of logs broke free and spilled onto the roadway. The driver jumped from the truck's cab and fled back to the west end of the bridge.

Frances Morris, the postal carrier, had passed mid-span when the twisting began. She fought to steer the car, as the wheels kept hitting the curb. Before safely reaching the east side, she saw and passed the car driven by Leonard Coatsworth as he neared the east tower.

Just after Coatsworth passed the east tower to the center span, the road tilted sideways and threw his Studebaker against the curb. One witness said the car was "bouncing around like a ball." Coatsworth tried to open the doors, but they were jammed. He then climbed through the open window and sprawled onto the road. He tried to coax Tubby from the back seat, but the frightened pet refused to move. Coatsworth began

toll plaza on the east end, a delivery van for the Rapid Transfer Company paid its toll and drove westward. Next came the *Tribune*'s Leonard Coatsworth, driving a dark green, two-door 1936 Studebaker sedan. Coatsworth was heading to the family's summer cottage at nearby Arletta, just six miles from the west end of the bridge, to close it up for the season. In the back seat rode his daughter's dog, a black spaniel named Tubby. Coatsworth paid the toll and drove his Studebaker onto the rippling span. Because of a phobia about driving in a closed car, his window was down, despite the chilly wind.

10:03 A.M. Suddenly, the roadway began a "lateral twisting motion." At first this movement was small. By 10:07 the motion became gigantic. The roadway tilted up to 28 feet on one side then the other at an angle of up to 35 degrees. Every five seconds the deck rose and fell violently with the twisting wave. When the galloping twist began, highway officials and the State Police quickly closed the bridge. They allowed only the press and Professor Farquharson onto the reeling span. After only

A view of the twisting roadway, looking west down the centerline from near the east tower. Coatsworth's car has slid to the left side of the roadway.
Gig Harbor Peninsula Historical Society, Howard Clifford, NB-041

staggering and crawling to the east tower, 150 yards away. He again thought of Tubby, and attempted to return to save the dog, but the violent motion threw him to the pavement again. He tried calling to the college student, Winfield Brown. The two haltingly struggled toward the east tower.

Near the tower, J.K. Smith and W.H. Kreiger, employees for the bridge painting contractor, Fisher Paint Company, had been cleaning part of the span. The bizarre movement and loud noises alarmed them. They saw Coatsworth further out on the center span, now staggering toward them, and they hurried off toward the toll plaza.

Coatsworth got to his feet and stumbled toward the Tacoma end of the span, some 480 yards distant. He reached the toll plaza, told the attendant about the dog in his car, then telephoned his office. The *Tribune* immediately dispatched photographer Howard Clifford and reporter Bert Brintnall. The news-

paper office also contacted freelance photographer James Bashford, who hastily headed for the bridge.

At the toll plaza, the attendant telephoned Barney Elliott of The Camera Shop in downtown Tacoma. Elliott and shop co-owner Harbine Monroe grabbed their movie cameras and several rolls of color film. They quickly drove to the bridge.

"As secure as the Narrows Bridge," proclaimed a large billboard for the Pacific National Bank along the 6th Avenue route between downtown Tacoma and the Narrows Bridge. The *Tribune*'s Howard Clifford and Bert Brintnall saw the sign on the south side of the road on their way to the Narrows. Clifford made a mental note to get a photo of the billboard on their return downtown.

Farquharson remained near the east tower shooting film footage. Walter Miles, a supervisor with the Pacific Bridge Company, also began taking motion pictures and telephoned

Clark Eldridge, telling him to come quickly because the bridge was "about to go."

Around 10:15 A.M. Clark Eldridge arrived at the bridge, seeing it "swaying wildly" with the bottom side visible as it tilted. Several people were struggling off the roadway at the east end. He joined Farquharson at the east tower to discuss the situation, and they returned to the east anchorage to warn people to stay off the span. Eldridge also telephoned the U.S. Coast Guard and the Northern Pacific Railroad to advise them of potential danger to marine or rail traffic passing beneath the bridge.

Around 10:30 A.M. More broken concrete chunks rolled about on the deck. Near the east tower, several lampposts wobbled loosely on their bases, and at the west end one light post fell over. A large chunk of concrete dropped out from a section of the center span and splashed into the Narrows. On the west end, two electricians, Robert Hall and March Wilson, backed their truck out onto the rolling bridge to rescue Ruby Jacox and Walter Hagen, who had abandoned their Rapid Transfer Company van.

At the east tower, Farquharson continued taking pictures. When his movie camera finished a roll of film, the professor ran quickly back to the toll plaza to get more. The *Tribune*'s Clifford, Bashford, and Brintnall arrived, and Clifford began taking photographs with his bulky Graflex camera. Then he and Brintnall ventured onto the span, joining Farquharson. There, too, was Barney Elliott from The Camera Shop, standing on the twisting bridge taking motion picture footage. Not long before, Elliott and Harbine Monroe had arrived and began shooting film footage from the south side of the bridge. Monroe remained there, while Elliott, as he later said, "got ambitious and started across the bridge," but he only made it to the east tower.[3]

For a short period, the wind subsided and the span steadied itself. Farquharson could see the dog in Coatsworth's automobile. He

decided to try to save Tubby. Hoping to drive the car to safety, he staggered out along the centerline and reached the vehicle. But now it seemed that the wind was blowing harder. He noted several lampposts on the west side falling over.

Suddenly, the bridge deck heaved violently. In the previous half-hour the suspended center-span structure had twisted 14 times per minute along its longitudinal axis (effectively the roadway centerline). During the brief lull, Farquharson had managed to walk the centerline, where the motion was at its least. But now the center span began twisting again, at 12 times a minute with a simultaneous sideways motion, with several peaks as it moved. Walking became extremely difficult.

The violent twisting of the bridge, accelerating to nearly the pull of gravity, shifted the

Around 10:30 A.M., the movement became dramatically more violent, now with lateral shifts. (Still frame from The Camera Shop's motion picture footage.)
Courtesy of The Camera Shop

Coatsworth's abandoned 1936 Studebaker. (Still frames from F.B. Farquharson's motion picture footage.)
UW Libraries, Special Collections, UW21424f

UW Libraries, Special Collections, UW21429

The bridge twists with several nodes. (Still photo from motion picture footage.)
Courtesy of The Camera Shop

car about on the deck. Alarmed, Farquharson attempted to grab Tubby. He thrust his left hand into the backseat, but the terrified dog bit the stranger's index finger. Farquharson struggled back to the east tower. Next, Howard Clifford ventured onto the center span to try to save Tubby, but after only a few yards had to turn back. He stopped at the tower near Brintnall and resumed taking photos.

Burt Farquharson also continued shooting photographs and motion pictures near the east tower. The suspended span now was twisting, Farquharson noted, at the astonishing rate of 20 times a minute. He walked haltingly to the centerline, straddled it, and squatted to better study the span's wild movements. Bystanders had the bizarre impression that Farquharson literally was "riding" Gertie's gallop. The professor still believed the bridge would settle down.

By now, word of Gertie's strange behavior had reached local radio stations, which broadcast reports that the bridge was "swaying dangerously." Curious residents began driving toward the Narrows. At the nearby College of Puget Sound, Beverlee Storkman happened to be near the campus newspaper room when word arrived that the bridge was is danger. "Most of us cut classes to go see it," she later recalled. "A group of friends and I piled into a car and raced over to the Narrows."

Already several dozen people had gathered to watch the spectacle, including Jeanette Taylor, a 19-year-old newlywed. Taylor and her husband had rushed to the Narrows from their home in Seahurst, north of Tacoma, after receiving a telephone call from friends.

"It looked like an undulating piece of chiffon," recalled Taylor. She stood with friends on the east approach span, looking out across the center of the twisting bridge. "I couldn't believe it. The power of it was overwhelming. Seeing the motion, my stomach was tight inside. It was frightening, watching the bridge buckling and wondering what was going to happen—you couldn't tell what was going to happen. I felt I had to get out of there. You could feel something terrible was happening. We were told to back away off the bridge, and we left right away. I never saw the bridge go down."[4]

About 10:55 A.M. A large chunk of concrete fell into the Narrows from a roadway section in the east half of the center span. Bystanders saw a thin band of daylight across the bridge deck where the piece fell out.

Some six minutes before the bridge began to fail, Farquharson tried once again to save Tubby. He started out from the east tower, but the motion had become too violent. He had to turn back. The professor stumbled back to the east tower just in time.

The *Atlanta*

The first—and last—vessel to cruise under the completed Narrows Bridge was the Coast Guard cutter *Atlanta*. The *Atlanta* had been the first ship to pass under the bridge when it opened on July 1, 1940. On November 7, the *Atlanta* was on a routine trip northward through the Narrows. Only moments before the first huge section of the roadway fell, the *Atlanta* passed beneath the rolling center span. Chunks of concrete hit the cutter's deck, but caused no damage. Lieutenant W.C. Hogan, commander of the vessel, watched closely as the bridge swayed and twisted above him. "It looked as though it would surely break up," Hogan told newspaper reporters later. When Gertie did fall into the Narrows, Hogan radioed the news to his Seattle headquarters, becoming the first to tell the world about the great disaster.

11:00 A.M. The extreme twisting waves of the roadway, magnified by the aerodynamic effect of wind on the girders of the bridge, began to rip the span apart. More chunks of concrete broke off "like popcorn," in the words of one witness, and fell into the chilly waters below. Massive steel girders twisted like rubber. Bolts sheered off and flew into the wind. Six light poles on the east end broke off as if they were matchsticks. Steel suspender cables snapped with a sound similar to gun shots, flying into the air "like fishing lines," as Farquharson later said. In the back of his mind now ran the thought, "The bridge's end has come at last. The dance of death has started."[5]

The strange sounds of the bridge's writhing filled the air. When the tie-down cables failed, the side spans began to work the main cables back and forth. The movement shifted the steel covers where the cables entered the anchorage, producing a metallic shrieking wail. By now, several hundred bystanders stood on the eastern shore. From the bluff, a workman on a pile driver repeatedly tooted his whistle to try to warn the approaching Coast Guard cutter *Atlanta*, which passed under the bridge. The shrill whistle blasts mixed eerily with the howling wind and the grinding and screeching of metal and concrete. The wild noise gave onlookers a sense of dread and impending calamity.

On the east side, Beverlee Storkman and her college classmates arrived and parked on a dead-end road. Her friends began walking, but, Storkman recalled, "It was really windy, and I wasn't going to get out of the car, but a girl friend yelled at me to come on. We walked up onto a little knoll. I saw the bridge, it twisted one last time."

Howard Clifford stood at the east tower, staring down into the lens of his camera, preparing to take another photo of Coatsworth's car and the galloping roadway. The bridge trembled. Clifford snapped the shutter, turned, and started to run. Reporter Bert Brintnall, who had been standing about 20 yards behind him, likewise turned and began to scramble toward the toll plaza.

11:02 A.M. A 600-foot section of roadway in the western half of the center span (the "Gig Harbor quarter point") broke free. With a thunderous roar the massive section wrenched from its cables in a cloud of concrete dust, flipped over, and plummeted 190 feet into Puget Sound. A mighty geyser of foam and spray shot upward 100 feet. Great sparks from shorting electrical wires also flew in the air.

Farquharson ran from the east tower toward the toll plaza, trying to cover the 1,100 feet of the east span as fast as his legs could carry him. Just in front of him, Howard Clifford sprinted, fell, and scrambled along the roadway. As they ran they heard suspender cables snapping with a sound like machine gun fire. Farquharson was retreating along the centerline, where he knew there was the least motion. Twice, the roadway dropped 60 feet, faster than the pull of gravity, then bounced upward, finally settling into a 30-foot deep sag. Farquharson fell, breaking one of his cameras. He rose to his knees, snapped another photograph, and again headed for the toll plaza.

In the bridge's mid part, successive deck sections rapidly fell out toward each tower. On the west half of the center span, the logging truck and the Rapid Transfer Company van went down with the crumbling road deck. On the east half, Coatsworth's car and Tubby followed the plunging roadway, tumbling end over end into the wind-swept Narrows.

11:10 A.M. It was over. The cold waters churned, eddied, and swirled as the heart of Galloping Gertie sank beneath the whitecaps, coming to rest on the bottom of the Narrows. By this time hundreds of cars were driving bumper-to-bumper to the bridge, making their way west on 6th Avenue from Tacoma and clogging side streets.

At 11:02 A.M., a 600-foot section of roadway from the west half of the center span breaks loose.
Washington State Archives

James Bashford's famous photo of Galloping Gertie's collapse.
Gig Harbor Peninsula Historical Society, Bashford 2786

The huge splash from the first massive section of the bridge hitting the water.
UW Libraries, Special Collections, UW21431

Foam and spray shoot up more than a hundred feet, while more of the suspended structure continues to break loose from Galloping Gertie.
UW Libraries, Special Collections, FAR019

A man running off the bridge, as the center span between the towers collapses and spray rises up to the right side of the roadway. Professor Farquharson follows closely behind this man. (Still frame from motion picture footage.)
UW Libraries, Special Collections, UW20731

The most spectacular failure in bridge engineering history was over. The visible remains of the world's third largest suspension bridge, the latest and most advanced in its sleek design, were a twisted tangle of steel and broken concrete.

LAST MAN

Newspapers at the time, and over the years since 1940, have given the title "Last Man on the Bridge" to four different people—Leonard Coatsworth, Howard Clifford, Barney Elliott, and Professor Farquharson. Some, too, have thought that the pedestrian, college student Winfield Brown, was due the title. In the months that immediately followed Galloping Gertie's collapse, Leonard Coatsworth was most widely heralded as the "Last Man on the Bridge." That may have been because newspapers across the country published the reporter's dramatic story of his flight off the doomed bridge and the loss of the car and Tubby.

In their departures from the bridge, Winfield Brown outran Coatsworth. Also at the time of the collapse, as Howard Clifford left the east tower area, reporter Bert Brintnall already was more than 20 yards ahead of him. Clifford, however, sprinted past photographer Barney Elliott, who only a moment before had begun to make his way to the toll plaza. Burt Farquharson, driven by the desire to record for engineering science the fate of the failing bridge, actually was the last man off Galloping Gertie.[6]

EYEWITNESS ACCOUNTS, NOVEMBER 7, 1940

Clark Eldridge—*Project Engineer, Washington State Toll Bridge Authority.*

I was in my office about a mile away when word came that the bridge was in trouble. At about 10 o'clock Mr. Walter Miles called from his office to come and look at the bridge, that it was about to go.

The center span was swaying wildly, it being possible first to see the entire bottom side as it swung into a semi-vertical position and then the entire roadway.

I observed that all traffic had been stopped and that several people were coming off the bridge from the easterly side span. I walked to tower No. 5 and out onto the main span to about the quarter point observing conditions. The main span was rolling wildly. The deck was tipping from the horizontal to an angle approaching 45 degrees. The entire main span appeared to be twisting about a neutral point at the center of the span in somewhat the manner of a corkscrew.

At tower No. 5, I met Professor Farquharson, who had his camera set up and was taking pictures. We remained there a few minutes and then decided to return to the east anchorage, warning people who were approaching to get off of the span.

At that time, it appeared that should the wind die down, the span would perhaps come to rest and I resolved that we would immediately proceed to install a system of cables from the piers to the roadway level in the main span to prevent any recurrence...

I was then informed that a panel of laterals in the center of the span had dropped out and a section of concrete slab had fallen. I immediately went to the south side of the view plaza. The bridge was still rolling badly. I returned to the toll plaza and from there observed the first section of steel fall out of the center. From then on successive sections towards each tower rapidly fell out.[7]

F.B. "Burt" Farquharson—*Professor of Engineering, University of Washington.*

I was the only person on the Narrows Bridge when it collapsed. When I arrived at about a quarter to ten o'clock, the bridge was moving in the familiar rippling motion we were studying and seeking to correct.

About a half hour later, it started a lateral twisting motion, in addition to the vertical wave. It had never done that before.

At least six lamp posts were snapped off while I watched. A few minutes later, I saw a side girder bulge out. But, though the bridge was bucking up at an angle of 45 degrees, I thought she would be able to fight it out. But that wasn't to be.

I saw the suspenders *[vertical cables]* snap off and a whole section caved in. The bridge dropped from under me. I fell and broke one of my cameras. The portion where I was had dropped 30 feet when the tension was released.

I kneeled on the roadway and stayed to complete the pictures.[8]

Winfield Brown—*A 25-year-old college student, Winfield Brown decided to walk onto Gertie shortly before 10:00 a.m. that morning. "I decided I'd like to get a little fun out of it," he later said. He paid the 10¢ toll and strolled onto the rolling bridge.*

After walking to the tower on the other side and back, I decided to cross again. It was swaying quite a bit. About the time I got to the center, the wind seemed to start blowing harder, all of a sudden. I was thrown flat. A car came up about that time. The driver *[Coatsworth]* got out, walking and crawling on the other side. We didn't have time for any conversation.

Time after time I was thrown completely over the railing. When I tried to get up, I was knocked flat again. Chunks of concrete were breaking up and rolling around. The knees were torn out of my pants, and my knees were cut and torn.

I don't know how long it took to get back. It seemed like a lifetime. During the worst parts, the bridge turned so far that I could see the Coast Guard boat in the water beneath.

As soon as I got off the bridge, I became sick. So I went to the home of a cousin and laid down for a while.

I've been on plenty of roller coasters, but the worst was nothing compared to this.

When I got back, I remembered the bridge man *[toll collector]* had said something about a dime each way. I mentioned it to him.

He said, 'Skip it.'[9]

Leonard Coatsworth—*News editor for the* Tacoma News Tribune.

I saw the Narrows bridge die today, and only by the grace of God, escaped dying with it.

I drove on the bridge and started across. In the car with me was my daughter's cocker spaniel, Tubby…

Just as I drove past the towers, the bridge began to sway violently from side to side. Before I realized it, the tilt became so violent that I lost control of the car…I jammed on the brakes and got out, only to be thrown onto my face against the curb.

Around me I could hear concrete cracking. I started back to the car to get the dog, but was thrown before I could reach it. The car

Onlookers view the ruined bridge.
Washington State Archives

itself began to slide from side to side on the roadway. I decided the bridge was breaking up and my only hope was to get back to shore.

On hands and knees most of the time, I crawled 500 yards or more to the towers…My breath was coming in gasps; my knees were raw and bleeding, my hands bruised and swollen from gripping the concrete curb…Toward the last, I risked rising to my feet and running a few yards at a time…Safely back at the toll plaza, I saw the bridge in its final collapse and saw my car plunge into the Narrows.

I saw Clark Eldridge, his face white as paper. If I feel badly, I thought, how must he feel?

With real tragedy, disaster and blasted dreams all around me, I believe that right at this minute what appalls me most is that within a few hours I must tell my daughter that her dog is dead, when I might have saved him.[10]

Ruby Jacox—*Several minutes before Leonard Coatsworth drove onto the Narrows Bridge, the next-to-last vehicle entering the reeling roadway from the east was the Rapid Transfer Company delivery truck. Inside were business partners Ruby Jacox and Walter Hagen, both age 45. It was Jacox's first trip over the Narrows Bridge. She had wanted to ride across on a windy day, "to get the thrill of feeling the structure sway."*

All of a sudden the bridge began to rock. We were afraid the truck would turn over, so we…jumped out. We could only crawl on our hands and knees and got about 10 feet away when the truck fell over.

We crawled along hanging onto the ridge of the center of the roadway. Just to keep our courage up we never stopped talking. Chunks of the concrete actually burst out of the bridge deck as it swayed, groaned and buckled. I fell dozens of times on the pavement.

I was ready to give up, but he (Mr. Hagen) just dragged me along by the shoulder. One of the lamp posts just did miss my head. Sometimes I was sure we'd never get off the bridge.

I kept thinking that this bridge was something that couldn't break. It had been inspected by government engineers. And experts had planned it so it would stand any strain.

[Jacox and Hagen were rescued by two painters who drove their truck out to retrieve them. Ruby Jacox suffered painful bruises on her knees, left hip, and ankles. She spent the night in a hospital recovering from "terrific nervous shock."][11]

Howard Clifford—*Photographer for the* Tacoma News Tribune, *28 years old at the time.*

I was on the Narrows Bridge when it broke in the middle and…I hope that I never again go through such a nerve racking experience.

The regular photographer was out on assignment. I was the back up, and they told me to grab a camera and go out there. But, they said, don't take any risks under any circumstances. I grabbed the only camera, an old Graflex, a large and cumbersome 4 x 5 reflex camera that you hold against your stomach and look down into the viewfinder.

When I arrived, the bridge had literally run amok, bouncing and twisting like a roller coaster. Working my way up to the tower with the greatest difficulty, I shot a few more films. Suddenly, the bridge seemed to sway and lurch more than ever, and I began shooting as fast as I could.

[Clifford then decided to walk onto the center span to try to save the dog, Tubby, in Leonard Coatsworth's car.] I probably wouldn't have gone out there, if it hadn't been for the dog. I liked dogs and had seen the Coatsworth's dog at a company picnic recently. Or, if I didn't have the camera, I probably wouldn't have gone out on the bridge. I got about 10 yards from the tower and stopped.

Taking another squint into the camera viewfinder, I saw the span buckle and start to break in the center. I pressed the camera trigger and started to run.

I tried to run up the yellow line in the center of the roadway, but found myself being bounced from one curb to the other and making no headway towards shore. I felt I could be tossed over the edge at any time. I was running in the air part of the time, because the bridge was moving faster than gravity. It dropped out from under me and then bounced back, knocking me down to my knees, banging the camera on the pavement.

Behind me I heard rumblings and explosive sounds which scared the daylights out of me. Having played football during my junior high and high school days, I tucked my camera under my arm and charging low got that added ounce of energy from somewhere which enabled me to make some headway toward the bridge entrance.

I was half-running, half-crawling. In a few minutes, which seemed like hours, I was up with my fellow photographer, who had got a considerable start, and we both made our way to the toll gate office, exhausted, but oh so thankful.

Returning to the *Tribune* office…within a very short time I was transmitting photos of the collapse of the Tacoma Narrows Bridge to the entire world. It was only then that I noted that my trousers were torn and my knees resembled raw hamburger. The next morning I looked even worse. I was bruised, black and blue from my hips to my feet the next day and for two weeks.

I don't think anything more exciting has ever happened to me.[12]

THE NEWS SPREADS

Leonard Coatsworth immediately telephoned the *Tribune* office in Tacoma from the toll plaza and dictated his account of Galloping Gertie's collapse. But he was in shock, and by the time he reached home, he had no memory of what he had said. Not until reading next morning's *Tribune* did he know what he had reported.

By the time Howard Clifford and Bert Brintnall left to get their photos and reports to the *Tribune*'s newsroom, there was no Tacoma Narrows Bridge poster on the bank billboard to photograph. Within an hour after the disaster, a crew of workmen covered the sign with plain white paper. But the story made it into the newspaper, and thus, into history.

By noon, radio newsmen also arrived at the Narrows. Carroll Foster from KIRO in Seattle reported from a small plane circling above the site, while other broadcasters milled about the east end of the bridge. They interviewed several people, including Senator Homer T. Bone. "This is the most astounding sight I have ever witnessed in my lifetime," exclaimed Bone.

In Olympia, the news shocked Governor Clarence D. Martin, who had considered the bridge one of the finest accomplishments of his two terms in office. He immediately cancelled all other business and headed north to the scene. Later, as he soberly viewed the ruined span, Martin sounded a positive note,

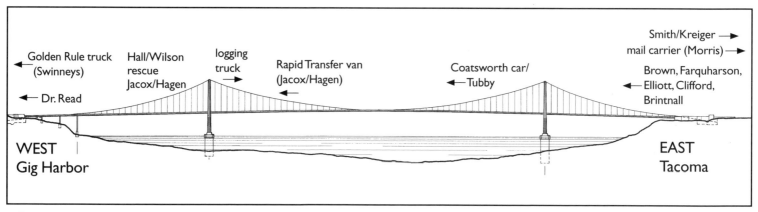

Before 11:02 A.M., November 7, 1940.
Kerry Darnall, Glen Lindeman, and Nancy Grunewald

assuring the public, "a greater bridge will rise from the wreckage."[12]

The press gave wide coverage to the spectacular event. So did national news magazines such as *Newsweek* and *Life*. And, in the years since 1940, various newspapers in the Puget Sound region have continued to print stories about the great collapse on November 7.

Notes

1. During the two weeks following the bridge failure, the *Tacoma News Tribune, Seattle Times,* and *Seattle Post-Intelligencer* provided in-depth coverage of the catastrophe—e.g., "Narrows Span Lies at Bottom of Sound," and "U. Engineer Describes Close Escape on Span," *Seattle Post-Intelligencer*, November 8, 1940; Howard Clifford, "Cameraman Was on Cracking Span," *Tacoma News Tribune*, November 8, 1940. Additional newspaper and other sources for the events and chronology of November 7, 1940, include: Advisory Board, *Failure of the Tacoma Narrows Bridge*, Appendix V; Petroski, *Engineers of Dreams*, 297–303; Farquharson, *Aerodynamic Stability*, 22–29; Joseph Gies, *Bridges and Men* (Garden City, New York: Doubleday, 1963), 243–53; Walter A. Averill, "Collapse of the Tacoma Narrows Bridge," *Pacific Builder and Engineer* 46 (December 1940): 20–27; N.A. Bowers, "Tacoma Narrows Bridge Wrecked by Wind," *Engineering News-Record* 125 (November 14, 1940): 647, 656–58; "Fall of the First Tacoma Narrows Bridge," *Washington Highway News* 12 (December 1964): 1–3; "Galloping Gertie," *Newsweek* 16 (November 15, 1940): 23–24; "Narrows Nightmare," *Time* (November 18, 1940); S. Ross, et al., "Tacoma Narrows, 1940," in *Construction Disasters* (New York: McGraw-Hill, 1984); "Galloping Gertie—Going—GONE!" *Seattle Times*, November 1, 1970; "Remember the Death of 'Galloping Gertie'?" *Seattle Times*, November 5, 1964; Gordon Schultz, "Galloping Gertie's Anniversary," *Seattle Post-Intelligencer*, November 7, 1965; Dick Stansfield, "'Galloping Gertie' Went into Narrows 25 Years Ago," *Daily Olympian*, November 6, 1967; Bart Ripp, "Cameraman Who Shot Galloping Gertie's Fall Dies," *Tacoma News Tribune*, July 24, 1997; Marcia Shannon, "'Galloping Gertie' Thrills Recalled," *Tacoma News Tribune*, November 3, 1974; "Where Were You when the Bridge Fell?" *Tacoma News Tribune*, November 1, 1990; letter, F.B. Farquharson to Theodore von Karman, December 1, 1940, Box 42, WTBA, WSA; Richard L. Swinney, telephone interview, October 2005.

2. Initially at the time of the collapse, Professor Farquharson estimated the bridge's tilt at 45 degrees, using lampposts along the bridge deck as his reference point. Later, after examining his motion picture footage, Farquharson realized that the lampposts were broken at their bases, exaggerating the angle. He revised the angle estimate later in *Aerodynamic Stability*, page 26.

3. Berney Elliott, interviewed by Enrique Cerna in the "Galloping Gertie" segment of the *Evening Magazine* program, KING-TV, Seattle, 1988.

4. Beverlee Storkman, telephone interview, September 17, 2006; Jeanette Taylor, telephone interview, September 2005.

5. Professor Farquharson is quoted from a 1964 interview with Dr. Tadaki Kawada, author of "Who Wrecked the Galloping Gertie? The Mystery of the Tacoma Narrows Bridge Disaster,"

an unpublished manuscript (1975; page 23). Kawada's initial text was written in Japanese, so Farquharson's words originally were translated from English to Japanese, and then back to English again in the later manuscript.

6. This conclusion coincides with the account of Ed Elliott, son of Barney Elliott, who claimed that his father always said, "Farquharson was the last one off the bridge." Ed Elliott interviews, 2003 and 2005.

7. Clark Eldridge in Advisory Board, *Failure of the Tacoma Narrows Bridge*, Appendix IV, 3–6; Eldridge, "An Autobiography," 44–49.

8. Farquharson, *Aerodynamic Stability*, 22–29; "Bridge Fell from Under Me, Professor Says," *Columbus Evening Dispatch*, November 8, 1940; "U. Engineer Describes Close Escape on Span," *Seattle Post-Intelligencer*, November 8, 1940; "Farquharson Dares Death on Crumbling Bridge," *University of Washington Daily*, November 8, 1940.

9. "Youth Who Crawled Off Narrows Bridge before Its Collapse Tells Experiences," *Tacoma Times*, November 9, 1940; "Span Mystery Man Had Big Thrill for Dime," *Tacoma News Tribune*, November 9, 1940; Margie Brown, telephone interview, October 9, 2006.

10. Gerry Coatsworth Holcomb, telephone interview, January 7, 2006, and various e-mails to the author, January–March 2006; "News Tribune Man Last on the Bridge," *Tacoma News Tribune*, November 8, 1940; "Man Trapped on Crumbling Bridge Tells Dramatic Story," *Seattle Post-Intelligencer*, November 8, 1940; "TNT Staffer Cheats Death in Bridge Collapse," *Tacoma News Tribune*, July 1, 1976; "Phobia Saved Newsman's Life," *Tacoma News Tribune*, November 7, 1975; "Leonard Coatsworth" folder, Box 42, WTBA, WSA.

11. "Tacoma Woman on Bridge Was Ready to 'Give Up,'" *Tacoma News Tribune*, November 9, 1940; "Couple Gets Extra Thrill: Crawl Off Bridge to Safety," *Bremerton Sun,* November 7, 1940.

12. Howard Clifford, interviews, October 2003 and September 2005; "Cameraman Was on Cracking Span," *Tacoma News Tribune*, November 8, 1940; Howard Clifford, "A Day to Remember; When the Tacoma Narrows Bridge Fell," TAPCO newsletter, undated copy in Tacoma Narrows Bridges History files, Gig Harbor Peninsula Historical Society; Howard Clifford, "Memories of Gertie," *Tacoma News Tribune*, November 1, 1990; "Lensman Recalls Gertie's Gallop," *Tacoma News Tribune*, November 7, 1975; "'Last Man on Bridge Recounts Gertie's Collapse," *Tacoma Sun*, November 2, 1980; Rob Carson, "A Run for His Life," *Tacoma News Tribune*, August 10, 2003.

13. Senator Bone quoted in November 7, 1940, radio interview tape-recorded by KIRO, Seattle; Martin quoted in *Seattle Times,* November 7, 1940.

A surviving portion of the
center-span roadway.
Washington State Archives

CHAPTER EIGHT

THE LESSONS OF FAILURE

GERTIE GALLOPS ON FILM

The striking photos and motion pictures taken during the Tacoma Narrows Bridge failure seem destined to fascinate people indefinitely. For more than six decades the images have been shown around the world, appearing in hundreds of publications, on movie and television screens, and today on Internet Web sites as well.

Five cameramen captured all or part of Galloping Gertie's spectacular collapse on motion picture film. Only one movie was shot from the west shore, taken by Dr. Jesse Read. The other four men took footage from the east, setting their cameras on the bridge itself or nearby on the shore. The footage proved to be extremely valuable for scientific study, illuminating the lessons gained from Galloping Gertie's failure.

Barney Elliott and Harbine Monroe, owners of The Camera Shop in Tacoma, had done contract filming for the Toll Bridge Authority in the course of the span's construction. Elliott, aware of concerns about Gertie's stability some weeks before the collapse, had arranged with the accountant at the toll plaza to telephone him, "if anything happens." When the call came on the morning of November 7, Elliott and Monroe grabbed their Bell & Howell 16-mm cameras and several packets of the new and expensive "Kodachrome" color film. They drove to the east end of the span and began filming; Elliott worked from the center and the right side (north) Monroe on the left (south).

Harbine Monroe captured the collapse sequence that became famous. Years later during an interview, Barney Elliott told television journalist Enrique Cerna, "My partner [Monroe] had his camera cranked up and just started running it. He didn't have the foggiest

The bridge in ruins, as seen from the east shore, November 1940.
Washington State Archive

notion whether he had it set right or anything else. He got the actual first part of the fall…I didn't have any idea we had as good a picture as we did."[1]

The Camera Shop sold the spectacular film to Paramount Studios, which converted the original color version to black and white and sent it to movie theaters around the world. Castle Films bought distribution rights and created thousands of 8-mm and 16-mm copies (with sound and without) for home viewing. The Castle Films version is easy to identify, with its dramatic title, "Disaster! The Greatest Camera Scoop of All Time." Also, because Paramount used a "sandwich" method to create the black and white version, all of that footage is reversed, or "flopped." Thus, in those film prints Coatsworth's car, which had slid into the left (south) lane, appears on the right side instead of the left.

The Camera Shop's film became a classic. Within a few weeks after the catastrophe, film attributed to Burt Farquharson also was made available for public purchase. Today, The Camera Shop offers for sale "The Collapse of the Tacoma Narrows Bridge" in color DVD and videotape versions, condensed to eight minutes and covering shots from construction to destruction.

Walter Miles of the Pacific Bridge Company took extensive color motion picture film during construction and afterward for the Washington Toll Bridge Authority. Miles also shot movies on November 7 and of the damage on succeeding days. He gave a copy of the footage to Professor Farquharson, and in December 1940 a copy also was sent to a federal investigating board. Several hours of film taken by Miles are in the WSDOT records at the Washington State Archives, but it appears no footage for the collapse itself is included.

The movies shot by Dr. Read, who had just safely crossed Galloping Gertie, were taken from the Gig Harbor side. He probably used an 8-mm camera (home movie size). Several blurry black and white prints from his film are held by the Gig Harbor Peninsula Historical Society, but the movie footage has disappeared.

As for the four cameramen shooting on the east side, unraveling the tangled issue of "who took what" is not easy. Today, The Camera Shop's owners claim they hold the only original film (16 mm) of the great collapse. Also, it appears that not all of Burt Farquharson's footage has survived. The loss may have occurred during the collapse, when the professor fell and broke one of his cameras, possibly exposing film, or it may be that footage was

A bridge inspector checks the damaged north cable.
Washington State Archives

accidentally ruined during the developing process at a film lab. Barney Elliott, a friend of Farquharson, gave a copy of The Camera Shop's footage to the professor, who combined this and film from Walter Miles with his own, then provided it to newsreel producers, government agencies, and the public. The University of Washington, which holds Farquharson's papers and records, claims to have original Farquharson footage and to own the copyright of the professor's films and still photographs.

Dramatic images of Galloping Gertie's collapse also were captured in still photography. It was common at the time to print photos from movie film. Those prints are easy to distinguish because the images are fuzzy. The best pictures came from the large format cameras of Howard Clifford, James Bashford, Professor Farquharson, and the Richards family studio, all of whom snapped memorable shots.

Farquharson's photographs are housed at the University of Washington and some are available through the University Libraries online exhibit. Howard Clifford and James Bashford, who snapped the most famous image of the collapsing bridge, donated their collections to the Gig Harbor Peninsula Historical Society. The Tacoma Public Library holds the Richards studio collection. In the immediate aftermath of the catastrophe, newspapers and other print media spread the images around the globe. Rarely did they run credit lines, so proper acknowledgements have largely been unassigned since 1940.[2]

THE DAMAGE

Cables

During the collapse, the main suspension cables were violently thrown side to side, twisted, and, in the mid-section, tossed 100 feet into the air. They slipped within their positions in the cable saddles atop each tower. In the bridge's mid span, a north cable band that slipped loose severed some 350 wires.

A slipping cable band severed approximately 350 wires in the north cable.
Washington State Archives

Other wires in the suspension cables also were severely stressed or distorted. The main cables were ruined and only worth salvaging as scrap metal. The collapse also broke many suspender cables. Some were gone or had fallen onto the approach spans, some sustained serious damage, and some remained unaltered. Their only value was as scrap metal. [3]

Towers

The main towers (West Tower #4 and East Tower #5), including the portal struts, were twisted and bent. Stress beyond the elastic limit of the metal resulted in buckling and permanent distortion. They, too, only had value as scrap metal.

The east tower (#5), bent more than 12 feet toward the east shore, November 11, 1940.
Washington State Archives

Girder damage.
Washington State Archives

The collapse stunned everyone, but especially the engineers. How could the most "modern" suspension bridge in the world, with the most advanced design, suffer catastrophic failure in a relatively light wind?

The State of Washington, insurance companies, and the U.S. government appointed boards of experts to investigate the collapse. Members of Washington's investigative board included: Russell Cone, resident engineer during construction of the Golden Gate Bridge and, following its completion, chief engineer until 1941; Francis Donaldson, a consulting engineer, and chief engineer in the early stages of the Grand Coulee Dam project; and Lief Sverdrup, senior partner of Sverdrup and Parcel, a St. Louis consulting engineering firm. Meanwhile, the insurance companies established a "Narrows Bridge Loss Committee."

The Federal Works Administration (FWA) appointed a three-member panel of top-ranking engineers: Glenn Woodruff, design engineer for the San Francisco-Oakland Bay Bridge; Theodore von Karman, aeronautical engineer and director of the David Guggenheim Aeronautical Institute at the California Institute of Technology; and Othmar Ammann, engineer involved in the design and construction of the George Washington, Triborough, and Bronx-Whitestone bridges in New York, and the Hell Gate and Golden Gate bridges. Woodruff, von Karman, and Ammann visited the Narrows Bridge site in December 1940 and met with government officials, as well as Professor Farquharson.

Some four months later, they submitted a report to the administrator of the FWA, John Carmody. Their March 1941 findings, which became known as the "Carmody Board" report, stated that the "random action of turbulent wind" caused the bridge to fail. This ambiguous explanation was the beginning of attempts to understand the complex phenomenon of wind-induced motion in suspension bridges. Three key points stood out:

Center Span and Side Spans

The concrete and steel center span (suspended structure) lying on the bottom of the Narrows was deemed a total loss. The center span's collapse, followed by the 30-foot dropping of the side spans, had caused other substantial damage. The plate girders and floor beams of the side spans were stressed and distorted, with some buckled beyond repair. The remainder of the broken concrete on the side spans needed to be removed, and the floor system had sections that were bent and overstressed. Again, the only value was as scrap metal.

Piers and Anchorages

Both the West Pier #4 and the East Pier #5 sustained no damage, except that the collapse of the center span had caused partial shearing of the rivets that attached the towers to the tops of the piers. The anchorages for the main cables were undamaged, although some of the concrete later had to be removed when the replacement bridge was built in order to allow room to spin new main cables.

Gertie Rests 30 Fathoms under the Sound

On November 28, 1940, the U.S. Navy's Hydrographic Office released locational information for the collapsed Tacoma Narrows Bridge— latitude 47:16:00 north, longitude 122:33:00 west, 30 fathoms deep.

---

1. The principal cause of the bridge's failure was its "excessive flexibility."
2. The solid plate girder and deck acted like a blunt, unstable airfoil, creating "drag" and "lift."
3. Aerodynamic forces were little understood, and engineers needed to test suspension designs using models in a wind tunnel.

"The fundamental weakness" of the Tacoma Narrows Bridge, said a summary article published in the *Engineering News-Record*, was its "great flexibility, vertically and in torsion." Several factors had contributed to the excessive flexibility:

1. The plate-girder suspended structure was too light.
2. The 8-foot-high stiffening girders were too shallow compared to the length of the side spans (1 to 350 ratio).
3. The side spans were too long compared with the length of the center span for such a flexible structure.

4. The cables were anchored at too great a distance from the side spans.
5. The width of the deck was extremely narrow compared with its center span length, an unprecedented ratio of 1 to 72.

The pivotal cause for the collapse, said the board, was the change from vertical waves to the destructive twisting, torsional motion. This event was associated with the cable band slippage on the north cable at mid-span. When the band slipped, the north cable was separated into two segments of unequal length. The resulting imbalance quickly affected the thin, flexible plate girders, which twisted easily. Once the unbalanced motion began, progressive failure followed. The board's most significant finding was obvious—the engineering community must study and better understand aerodynamics in designing long-span suspension bridges.

Meanwhile, Professor Farquharson continued his wind tunnel tests. He concluded that

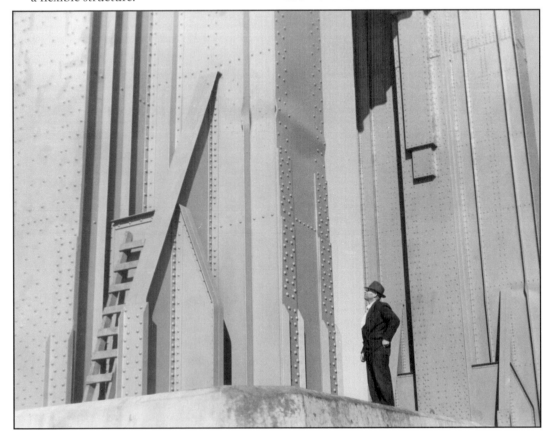

A man stands atop the east pier, dwarfed by the tower base. Plates just behind and in front of him show buckling from the stress of the November 7 collapse.
Washington State Archives

Clark Eldridge and Leon Moisseiff touring the bridge a week after the collapse, November 15, 1940.

Tacoma News Tribune

the bridge's lightness and "cumulative effect of undamped rhythmic forces" had produced "intense resonant oscillation" that caused the bridge to fail. The professor was on the right track. His investigations were beginning to shed light on the complex phenomenon that caused Gertie's collapse. Still, much more work remained before the root cause of the failure would be uncovered.[4]

Leon Moisseiff's design, while pushing the boundaries of suspension bridge engineering practice, fully met the requirements of accepted theory at the time. Immediately after the bridge's failure, the media contacted him for his reaction. He replied simply that he was "completely at a loss to explain the collapse."

Exactly a week after the Narrows disaster, Moisseiff visited the ruined bridge with Clark Eldridge on a rainy Thursday. A photographer for the *Tacoma News Tribune* snapped an image of the two men as they stood on the east approach span. The photograph reveals a quiet drama. With the bridge's east tower looming eerily in the mist behind them,

Eldridge is gesturing with his right hand as he explains the tragic events, while a grim-faced Moisseiff, his jaw set, stares blankly past the camera's eye away from the bridge.[5]

"BLIND SPOT"—DESIGN LESSONS OF GERTIE'S FAILURE

At the time the bridge failed, the small community of suspension bridge engineers believed that lighter and narrower bridges were theoretically and functionally sound. Few people were designing these huge civil works projects. In general, leading designers like David Steinman, Othmar Ammann, and Leon Moisseiff determined the direction of the profession. The great bridges were extremely expensive and presented immensely complicated engineering and construction problems. The work also was sharply limited by government oversight and constant public scrutiny. Under these circumstances, a handful of talented engineers had risen to preeminence, but they had a blind spot.

Sagging east side span.
Washington State Archives

This blind spot was the root of the problem, according to bridge historian David P. Billington. At that time among suspension bridge engineers, says Billington, "there seemed to be almost no recognition that wind created vertical movement at all."[6]

The best suspension bridge designers in the 1930s believed that earlier failures had occurred because of heavy traffic loading, poor workmanship, and wind, but wind was not seen as being particularly important. The numerous wind-caused suspension bridge failures in the 19th century seemed to have little relevance to contemporary bridges, which were much bigger and heavier. Furthermore, modern bridges were designed according to the Deflection Theory.

Engineers viewed stiffening trusses as important to prevent sideways movement (lateral deflection) of the cables and roadway. They believed sideways motion resulted from temperature changes and static wind pressure (i.e., the wind "pushing" a suspended structure sideways). However, they thought little of the dynamic effects of wind. Traffic loads—when distributed unequally in the lanes on either side of a span (which they almost always are)—could, of course, induce torsion, and this was accounted for in the design. Restraint against torsion was required to resist structural deformation when traffic was concentrated on one side of a road deck, but not to counter the wind.

This trend ran counter to the lessons of previous times. Early suspension bridge failures occurred because the light spans with their very flexible decks were vulnerable to dynamic wind forces. Thus, in the late 19th century engineers had moved toward constructing stiffer and heavier suspension bridges. John Roebling consciously designed the 1883 Brooklyn Bridge so that it would be stable against the stresses of wind. By the early 20th century, however, Roebling's "historical perspective seemed to have been replaced by a visual preference unrelated to structural

engineering," according to historian David Billington.[7]

Just four months after Galloping Gertie failed, a civil engineering professor at Columbia University, J.K. Finch, published an article in the *Engineering News-Record* that summarized more than a century of suspension bridge failures. In the essay—titled "Wind Failures of Suspension Bridges, or Evolution and Decay of the Stiffening Truss"—Finch reminded engineers of some important history as he reviewed the record of spans that had suffered from aerodynamic instability. Finch declared, "These long-forgotten difficulties with early suspension bridges, clearly show that while to modern engineers, the gyrations of the Tacoma bridge constituted something entirely new and strange, they were not new—they had simply been forgotten."[8]

An entire generation of suspension bridge engineers had overlooked the lessons of the 19th century. The last major suspension bridge failure had occurred five decades earlier, when the Niagara-Clifton Bridge fell in 1889. Furthermore, in the 1930s aerodynamic forces were not well understood at all. As their experience with leading-edge suspension bridge designs gave engineers new knowledge, they had failed to relate it to the dynamic effects of wind forces.

END OF AN ERA

The collapse of Galloping Gertie revealed the limitations of relying on the Deflection Theory alone. Now, engineers could see that suspension bridges needed to be further stiffened. Thus, after November 7, 1940, engineers rushed to evaluate other suspension bridges and take measures to minimize potentially dangerous oscillations. They added stiffening to the trusses of the Golden Gate Bridge and to the slender plate girders of the Thousand Islands, Deer Isle, and Bronx-Whitestone bridges.

The Narrows disaster, for all practical purposes, ended Leon Moisseiff's career.

Interestingly enough, in association with Frank Masters he had submitted a plan earlier in 1940 for bridging the Straits of Mackinac; confident in his design of the Narrows structure, the proposal called for a 4,600-foot span with identical features. After Tacoma, Moisseiff received support from his colleagues, but life for him was not the same. Professional work in the remainder of his career, which soon ended with his death in 1943, was largely limited to consultation on the causes behind the Narrows Bridge failure.

Most importantly, the collapse abruptly altered bridge engineering theory and practice, and the trend in designing increasingly flexible, light, and slender suspension spans. Othmar Ammann said of the collapse, "Regrettable as the Tacoma Narrows Bridge failure and other recent experiences are, they have given us invaluable information and have brought us closer to the safe and economical design of suspension bridges against wind action."

Clark Eldridge knew as soon as anyone else what the broader implications of the Narrows Bridge catastrophe were. He stated to Theodore Condron on November 16, only nine days after the event, "The one outstanding fact about the whole thing seems to be that hereafter in the design of these structures, full consideration will have to be given…to the principles of aero-dynamics. I have no doubt that the disaster will mark a great step in our knowledge of design."[9]

SUSPENSION BRIDGE DESIGN SINCE 1940

The end of the 1950s witnessed the construction of two of the greatest suspension bridges in the world, designed by a pair of the 20th century's finest bridge engineers. The Mackinac Bridge, which opened in November 1957 in Michigan, was David B. Steinman's crowning achievement. New York's Verrazano-Narrows Bridge, designed by Othmar Ammann, was 10 years in the making and opened in November 1964. Both of these monumental spans directly benefited from the legacy of the failed 1940 and the successful 1950 Tacoma Narrows bridges. For more than six decades now, engineers have created suspension bridges that are stiffened against torsional motion, or are aerodynamically streamlined, or both.

Buckled steel beams on the bottom side of the suspended structure.
Washington State Archives

The Narrows failure revealed for the first time the limitations of the Deflection Theory. Since the Tacoma disaster, aerodynamic stability analysis has come to supplement the theory, but not replace it; the Deflection Theory remains an integral part of suspension bridge engineering. Today, the theory's principles serve as a model for the complex analytical methods (such as "Finite Element" computer programs) used by engineers to calculate stresses in suspension cable and suspended structure systems.

Wind tunnel testing for aerodynamic effects on bridges is now commonplace. In fact, the U.S. government requires that all bridges built with federal funds have the preliminary design subjected to wind tunnel analysis using a three-dimensional model. And, since the 1990s, advances in computer simulation technology and high-speed processing have enabled engineers to use advanced modeling software to conduct thorough aerodynamic analyses of the structures they design.

Why Did Galloping Gertie Collapse?

Engineers have studied the collapse of the 1940 Tacoma Narrows Bridge for more than six decades. The experts, however, have disagreed on at least some aspects of the explanation. Only in the last few years has a definitive description that meets almost unanimous agreement been reached. In fact, the recent work of Danish bridge aerodynamicist Allan Larsen is a fascinating story in and of itself, because it has been generally accepted as *the* explanation for the collapse more than six decades *after* the event.

Careful research conducted since the Carmody Board's report of March 1941 revealed that the original explanation given for the failure was only partly correct. In the immediate wake of the disaster, the Advisory Board on the Investigation of Suspension Bridges assembled the nation's leading engineers in an exhaustive 12-year effort (from 1942 to 1954)

to understand and incorporate design measures to counter the aerodynamic effects of wind on suspension bridges. The 30-member board included F.B. Farquharson, Theodore von Karman, Othmar Ammann, Leon Moisseiff, and Charles Andrew among the leading investigators. Their work also assisted Andrew and the WTBA in developing an aerodynamically stable design for the second Tacoma Narrows Bridge (1950).

This research included wind tunnel testing and the development of a mathematical theory to explain aerodynamic phenomena and its affect on suspension bridges. This led to an understanding of "flutter," which actually was well-known in aeronautics studies. Von Karman had advocated—unsuccessfully—with the Carmody Board in 1941 that flutter was the probable cause of the Tacoma bridge's demise, and generally a key threat to suspension bridges. Although bridge builders did abandon the solid plate girder in favor of deep, open, and rigid trusses (and, beginning in the 1960s, the revolutionary box girder of Great Britain's Severn suspension bridge), the exact nature of Gertie's demise was largely a question of academic and historical interest. Other plate girder suspension bridges, particularly the Bronx-Whitestone Bridge in New York, continued to exhibit aerodynamic instability, despite several remedial measures adopted over the years.[10]

Only recently has a comprehensive explanation been developed. In 2000, Allan Larsen published his research, benefiting from more advanced computer modeling and mathematical theories. The primary explanation of Galloping Gertie's failure is described as "torsional flutter," a complicated series of events:

1. In general, the 1940 Narrows Bridge had relatively little resistance to torsional (twisting) forces. That was because it had such a small depth-to-width ratio, 1 to 72, and Gertie's long, narrow, and shallow plate girder suspended structure, plus its "H" cross-section, made the bridge extremely flexible.

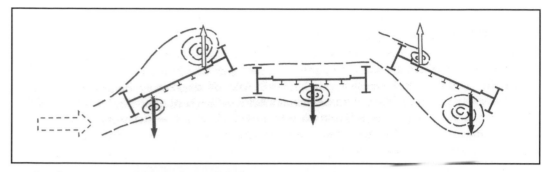

2. On the morning of November 7, 1940, shortly after 10:00 a.m., a critical event occurred. The cable band at mid-span on the north cable slipped. This allowed the cable to separate into two unequal segments, contributing to the change from vertical (up-and-down) to torsional (twisting) movement of the bridge deck. The change led Larsen to the core of the explanation for the collapse of the bridge.

3. "Torsional flutter" is a complex mechanism. "Flutter" is a self-induced harmonic vibration pattern. The instability can grow into very large vibrations. When the movement changed from vertical to torsional oscillation, the structure absorbed more wind energy. The deck's twisting motion began to control the wind vortex so the two were synchronized and the structure's twisting movements became self-generating. In other words, the forces acting on the bridge were no longer caused by wind alone; the deck's own motion helped produce the forces. Engineers call this "self-excited" motion.

4. "Self-excited" torsional flutter occurred in the Narrows Bridge as follows:
(a) The wind separated as it struck the upwind side of Galloping Gertie's 8-foot solid plate girder. A small amount of twisting occurred in the bridge deck, because even steel is elastic and changes form under high stress. Also, static wind forces "pushed" the plate girders a bit to one side, which induced a slight change in angle.
(b) The twist in the plate girder caused the wind flow separation to increase.

This formed a vortex, or swirling wind force, which further lifted and twisted the suspended structure as the vortex moved about the deck.
(c) The suspended structure resisted this lifting and twisting because of a natural tendency to return to its previous position. As it returned, its speed and direction matched the lifting force. In other words, the suspended structure moved "in phase" with the vortex. Another vortex then traveled from the leading edge of the plate girder beneath the deck, reinforcing that motion and pulling the deck down. This produced a "lock-on," or continuing, event.

5. The external force of the wind alone was not sufficient to cause the severe twisting that led to the failure. This is the key to "torsional flutter"; it is significant that the two types of instability, vortex shedding and torsional flutter, *both* occurred at relatively low wind speeds. Usually, vortex shedding occurs at relatively low wind speeds, like 25 to 35 mph, and torsional flutter at high wind speeds, such as at 100 mph. Because of Gertie's design and relatively weak resistance to torsional forces, the vortex shedding instability progressed directly to "torsional flutter," despite the low wind speed.

Now the bridge was beyond its natural ability to "damp out" the motion. Once the twisting movements began, they controlled the vortex forces. The torsional motion began small and built upon its own self-induced energy. In other words, Galloping Gertie's twisting induced more twisting, then greater and greater twisting, increasing beyond the structure's ability to resist. Failure resulted.[11]

What If...?

What if Clark Eldridge's original design for the Tacoma Narrows Bridge had been adopted instead of Leon Moisseiff's? Would it have blown down on November 7, 1940? Probably not—the bridge would still be there. This is the opinion of leading bridge engineers who have carefully studied Eldridge's design, with its 25-foot-deep stiffening truss. "I believe without a doubt," said one senior structural engineer, "that the bridge would have been aerodynamically stable for the wind speeds that destroyed Galloping Gertie."

Notes

1. Barney Elliott, interviewed by Enrique Cerna in the "Galloping Gertie" segment of the *Evening Magazine* program, KING-TV, Seattle, 1988.

2. Ed and Darcie Elliott, interview, September 2005; Kerry Webster, "Famous Collapse Footage Was Shot in Vivid Color, *Tacoma News Tribune*, November 2, 1975; Bart Ripp, "Cameraman Who Shot Galloping Gertie's Fall Dies," *Tacoma News Tribune*, July 24, 1997; Marcia Shannon, "'Galloping Gertie' Thrills Recalled, *Tacoma News Tribune*, November 3, 1974; "Film Showing Collapse of Tacoma Span Available," *Engineering News-Record* 125 (December 5, 1940): 733; "Movie Camera Record of Big Span's Collapse" [Farquharson stills], *Seattle Post-Intelligencer*, November 8, 1940; Walter Averill, "Collapse of the Tacoma Narrows Bridge," *Pacific Builder and Engineer* 46 (December 1940): 20–27 [Averill's article includes subsection, "Chronological Record Caught by Camera," on pp. 22–23]; Advisory Board, *Failure of the Tacoma Narrows Bridge*, section B, 18, 31 [regarding Miles film].

3. Damage to the bridge is described in the following: Farquharson, *Aerodynamic Stability*, 29–31; Advisory Board, *Failure of the Tacoma Narrows Bridge*, I-III; "Details of Damage to Tacoma Narrows Bridge," *Engineering News-Record* 125 (November 14, 1940): 647; "Details of Damage to Tacoma Narrows Bridge," *Engineering News-Record* 125 (November 21, 1940): 674–76; "Details of Damage to Tacoma Narrows Bridge," *Engineering News-Record* 125 (November 28, 1940): 720; "Dynamic Stability of Suspension Bridges" (editorial), *Pacific Builder and Engineer* 46 (December 1940): 1; "Dynamic Wind Destruction" (editorial), *Engineering News-Record* 125 (November 21, 1940): 672–73; "Tacoma Narrows Bridge Dismantling Recommended by Engineers," *Engineering News-Record* 126 (March 27, 1941): 453; "Field Books," 1940, Box 47, WTBA, WSA.

4. Advisory Board, *Failure of Tacoma Narrows Bridge*; "Another Consultant Board Named for Tacoma Span," *Engineering News-Record* 125 (December 5, 1940): 735; "Board Named to Study Tacoma Bridge Collapse," *Engineering News-Record* 125 (November 28, 1940): 725; "Why the Tacoma Narrows Bridge Failed," *Engineering News-Record* 126 (May 8, 1941): 75–79; "Action of 'Karman Vortices'" (Blake D. Mills letter to editor), *Engineering News-Record* 125 (December 19, 1940): 808; "Tacoma Bridge Report Released by PWA Board of Consultants," *Engineering News-Record* 126 (March 27, 1941): 589.

5. B.W. Brintnall, "Strong New Span, Engineer's Promise," *Tacoma News Tribune*, November 15, 1940; "Span Model Withstood Gale Says Bewildered Designer," *Seattle Times*, November 14, 1940.

6. David P. Billington, "History and Esthetics in Suspension Bridges," *Journal of the Structural Division, ASCE* (August 1977): 1655–72.

7. *Ibid.*

8. J.K. Finch, "Wind Failures of Suspension Bridges, or Evolution and Decay of the Stiffening Truss," *Engineering News-Record* 126 (March 13, 1941): 74–79. See also, J.H. Cissell, "Stiffness as a Factor in Long Span Suspension Bridge Design," *Roads and Streets* 84 (April 1941): 64, 67–68.

9. "Memoir of Leon Solomon Moisseiff" (prepared by O.H. Ammann and Fredrick Lienhard), *Transactions of the American Society of Engineers* 111 (1946): 1509–12; David Plowden, *Bridges: The Spans of North America* (New York: W.W. Norton, 1984), 119; letter, Clark Eldridge to Theodore Condron, November 16, 1940, Box 102, von Karman Papers, Caltech.

10. See Richard Scott's *In the Wake of Tacoma* for a thorough discussion of these developments.

11. The author gratefully acknowledges assistance from the following individuals: Richard Scott, various e-mails 2003–2005, and *In the Wake of Tacoma*, 89–93 and 345–54; Allan Larsen, COWI Architectural Services, Denmark, correspondence, June 7, 2004; Allan Larsen, "Aerodynamics of the Tacoma Narrows Bridge—60 Years Later," *Structural Engineering International* 4 (2000): 243–48; Tim Moore, Senior Structural Engineer, Washington State Department of Transportation.

View from the east; concrete sections are being
removed from the roadway.
Washington State Archives

CHAPTER NINE

A CHALLENGING BEGINNING: 1940–1948

SCANDAL: WHO WAS TO BLAME?

"U.S. Money-Lenders Blamed by Engineers for Span Crash," read the headline in the *Tacoma Times* on November 9, 1940, two days after the collapse of Galloping Gertie. When reporters asked lead project engineer Clark Eldridge to explain why the Narrows Bridge failed, he did not hold back. He was angry.

Eldridge told the newspapers: "The men who held the purse-strings were the whip-crackers on the entire project. We had a tried-and-true conventional bridge design. We were told we couldn't have the necessary money without using plans furnished by an eastern firm of engineers, chosen by the money-lenders." He and other state engineers had protested Leon Moisseiff's design and its 8-foot solid girders, which Eldridge called "sails."

For a week, stories relating to the collapse dominated the front pages of regional newspapers and were picked up by the Associated Press, which sent them across the country. Congressman John Coffee, representative from the Tacoma area, blasted federal agencies for under-funding the bridge.[1]

Authorities from the Public Works Administration and Reconstruction Finance Corporation, who had financed the bridge, scrambled to defend themselves. They promptly disputed charges of improper actions, and said they knew nothing about a problem with the design. Their strong reaction gave the impression of being overly defensive, as if they in fact had something to hide.

The surviving PWA records at the National Archives in Washington, D.C., include a list of 19 queries, labeled "Questions on Tacoma Bridge," with answers prepared for the agency head, dated December 4, 1940. Apparently these were prepared for J.J. Madigan, the Acting Commissioner of Public Works, to respond to inquiries from the press. The more general procedural questions also became incorporated into the agency's reply to the Carmody Board's request for information, later published in the board's report. The answers explicitly addressed the issues raised by Clark Eldridge in the newspapers, and consistently used the words "there is no record of any such report." Regarding the matter of the original design, the PWA position was carefully worded: "P.W.A. never heard of 'Clark Eldridge's design' as such."[2]

The agency's position was repeated a week later. In a telephone conversation on December 10, 1940, a reporter for the Scripps-Howard newspaper syndicate questioned a PWA official, Harry M. Brown, about the bridge failure. Despite persistence by the reporter, Brown denied there had been any significant changes in the bridge's design.

Workmen remove concrete, October 29, 1942. *Washington State Archives*

There was "no record here," said Brown, of any protests by Washington State's bridge engineers, and the PWA had not rejected the original design. Carefully choosing words, Brown said, "There was no plan given to us labeled the Eldridge design." The reporter pressed the issue again, to which Brown carefully replied, "Not to my knowledge."[3]

However, both the PWA and the RFC knew of problems with the bridge. David L. Glenn, the PWA's field engineer on site in Tacoma, submitted regular progress reports on the bridge's construction, until its collapse. In fact, Glenn sent his superiors a report warning of faults in the design and he refused to recommend acceptance of the structure. The other federal office involved, the RFC, certainly knew of the instability problems and all the remedial measures attempted. The RFC's field engineer on site, James Roper, sent weekly progress reports to his boss, Morton McCartney, chief of the Self-Liquidating Section at the RFC in Washington, D.C., and

mentioned the bridge's unusual vibrations. Moreover, Clark Eldridge had sent reports to the RFC's consulting engineer, Theodore Condron.

One of the PWA's own engineers soon broke the truth to the newspapers and the public. A month later, the scandal again made headlines: "Bridge Exposé Breaks" proclaimed the *Tacoma Times* on January 11, 1941. This time, it became public that PWA engineer David Glenn had refused to approve the bridge when it was completed in July 1940. But the PWA accepted the bridge, as did the Washington Toll Bridge Authority. The PWA fired Glenn two weeks after the story made headlines—newspapers reported that he had been "relieved" of his position on January 25, 1941.[4]

As the Carmody Board neared the end of its investigation, the WTBA saw no need to continue its relationship with Leon Moisseiff. On February 10, 1941, it issued the consulting engineer his last payment for

Dismantling the main suspension cables, February 1943.
Washington State Archives

services. A month later in March 1941, the Carmody Board announced its findings but refused to blame any one person. The entire engineering profession was responsible, said the experts, exonerating Leon Moisseiff. After November 7, 1940, however, Moisseiff's services were not in high demand. The Carmody Board's report contained a statement by the Acting Commissioner of Public Works, J.J. Madigan, explaining the selection of consulting engineers for the Narrows project. It included the following statement: "In no instance did this Administration nominate, or express any preference for any particular individual, group or firm."[5]

Clark Eldridge knew otherwise. In local newspaper accounts published on November 9, 1940, he bluntly blamed Moisseiff and federal authorities. Eldridge declared that Moisseiff and the consulting firm of Moran & Proctor "associated themselves to secure the commission to design the Tacoma bridge. They went to Washington, called on the Public Works Administration and informed them that they could design a structure here that could be built for not more than $7,000,000. So when Mr. Murrow appeared asking for $11,000,000, our estimate, he was told $7,000,000 was all they would approve. They suggested that he confer with Mr. Moisseiff and Moran & Proctor. This he did, ending up employing them to direct a new design."[6]

One month after the Carmody Board's report became public, Clark Eldridge decided he needed a career change. In April 1941, he resigned from the State Highway Department and took a job with the U.S. Navy on Guam in the South Pacific.

IMPACT ON TACOMA AND THE PENINSULA

For Tacoma and the Peninsula, the collapse of the Narrows Bridge was a calamity. The U.S. military lost a vital link between the Bremerton Naval Yard and the Army's installations at McChord Field and Fort Lewis for the duration of World War II. Merchants on both sides of the Narrows lost income in the retail trade between Pierce and Kitsap counties. For many Peninsula residents the Narrows Bridge immediately had become a lifeline, connecting their rural area to commercial centers in Tacoma and even Seattle.

Disappointment ran high in Tacoma, too. "Bridge Price too Low," wrote the *Tacoma Times*, while stating that the Narrows Bridge had been lost because it was "designed to fit a price." The newspaper blasted federal authorities for "an experiment in skimpiness." Just months earlier, the local media had praised Moisseiff's design as "graceful." Now they dismissed it as "skinny."

The federal officials and eastern politicians who had restricted funds for the Narrows Bridge had underestimated the region's overall need for the span and its value to the people in surrounding communities. Gertie's popularity had proved them wrong. In its four months

Dismantling a tower, May 1943.
Washington State Archives

of use, the toll revenues fully justified a bigger, better, and more expensive bridge. South Puget Sound could have afforded a four-lane span costing $10 million, which would have been strong, sound, safe—and still standing at the Narrows.

Charles Andrew was diplomatic, but also forceful. In a 1945 article in the *Engineering News-Record* entitled "Redesign of the Tacoma Narrows Bridge," Andrew wrote: "Traffic over the original bridge, during its brief existence, indicated that a structure of considerably greater cost would have been justified."[7]

Restoring ferry service for cross-Narrows travelers received top priority. In 1938, the State had purchased Mitchell Skansie's Washington Navigation Company. Immediately after the collapse, officials moved quickly. By 6:00 A.M. the very next morning, two ferries began steaming over the route made obsolete just four months before. By the end of the decade, when the 1950 Narrows Bridge was completed, commuters were ready for the new span.[8]

THE MONEY—SCANDAL, SQUABBLES, AND LAWSUITS

Galloping Gertie left a tangle of financial issues for State officials and the Sound country to unravel. From insurance litigation to larceny, from salvage to the future funding of

a replacement bridge, the money side of the collapse's aftermath became a long chain of frustrating events.

The bridge carried insurance spread among 22 different companies, with a total insured value of $5.2 million, or 80 percent of its full cost. Just prior to the bridge's failure, revenues had been so much higher than expected that the WTBA had considered dropping insurance on the span. When the structure collapsed, the lives of some insurance men suddenly became very interesting. Hallett R. French certainly became excited at the news of Galloping Gertie's demise. French had pocketed premiums on one of the State's policies and never reported the transaction to his company. He felt sure that he would never be found out. He was, of course, and went to prison for his bungled criminal effort. If the bridge had not collapsed, or if it had failed two or three weeks later, the WTBA might have cancelled its insurance policies, and French never would have been caught.

Meanwhile, the State and the insurance companies became embroiled in settlement squabbles. On June 2, 1941, the insurance underwriters filed their report. They believed that the piers, cables, and towers could be salvaged and reused, and offered the State a settlement of $1.8 million. Three weeks later, on June 26, 1941, the State filed its claim. Except for the piers, said the State, the bridge

TRAFFIC, 1930–1950.

Year	Total Annual Vehicles	Average Daily Vehicles
1930 (ferry)	171,993	471
1935 (ferry)	165,724	454
1939 (ferry)	205,842	564
1940 (bridge, July 1–Nov. 7)	265,748 (avg. 66,437/mo.)	2,044
1940 (ferry)	144,587	396
1945 (ferry)	480,009	1,315
1950 (ferry, Jan. 1–Oct. 13)	593,871 (avg. 59,387/mo.)	1,627
1950 (bridge, Oct. 14–Dec. 31)	280,464 (avg. 93,488/mo.)	3,550

was virtually a total loss, estimated at almost $4.3 million.

When the legal dust settled in August 1941, the two sides agreed on a settlement of $4 million. Considering the fact that the still usable piers alone had cost $2.4 million of the bridge's total $6.4 million budget, the settlement amounted to a resounding victory for the State.

The Washington Toll Bridge Authority next faced the task of replacing the Narrows Bridge. Unfortunately, the legal and insurance hassles had taken more than nine months to resolve, and World War II would intervene

before funding could be secured and a new bridge started.

SALVAGING THE 1940 BRIDGE

After most of the center span fell into the Narrows, the towers, main cables, side spans, and anchorages remained. The mangled superstructure was a serious hazard to boats navigating the Narrows. To minimize risks, dismantling began almost immediately. The salvage process on the ruined bridge proved as intricate and dangerous as its construction.

Salvage work on the steel suspended structure. *Washington State Archives*

The first stage of salvage, removing the dangling sections of the bridge's floor system, started in December 1940 and ended five months later. In March 1941, estimates for the continuing salvage operation looked gloomy. Experts calculated the scrap metal value of the towers, remaining plate girders, and other steel parts was $206,000. It would cost three times that amount, an estimated $636,000, to dismantle the bridge.

Dismantling officially began on September 8, 1941, and ended in June 1943. With America's entrance into World War II in December 1941, hopes rose that a steel shortage would bring a better price, and perhaps a profit, but that optimistic vision faded. By June 1943, workers had removed more than 4,000 tons of steel from the towers and 3,000 tons from the main cables and suspended structure. The WTBA paid nearly $646,661 for the effort. The 7,000 tons of steel was sold for scrap; in return the State received a meager $295,726. In the end, salvage operations posted a net loss of $350,935.[9]

The fate of the salvaged materials from Galloping Gertie is an interesting story in itself. America's growing armaments industry and war machine acquired most of the steel from Gertie. The State Highway Department also purchased some 413 tons of structural steel for use in new bridges. And some of the east approach span went into erecting a bridge on the strategically important Alcan Highway in Canada that connected Alaska with the lower 48 states.

Gertie's main cable wire only had value as scrap steel. Nevertheless, it spun one of the fascinating myths about the 1940 bridge. A story claims that the salvaged wire was used in Canada's Peace River Bridge, which also eventually failed, inferring a mysterious connection between the two disasters. In fact, salvaged parts from Galloping Gertie were used in constructing a different bridge, across Canada's Liard River, and that span did not fail.

The two Alcan Highway bridges—the Peace River Bridge and the Liard River Bridge—were completed in 1943 and 1944, respectively. Two sections from the Tacoma bridge's east approach were used in the Liard span. The Liard Bridge never collapsed and stands today where it was built. However, the Peace River Bridge (with no Galloping Gertie materials) later failed, but for reasons very different from the causes of the Narrows collapse.

The Peace River suspension bridge was built by the same company that constructed the main suspension cables at the Narrows, John A. Roebling's Sons Company. A variety of serious problems plagued the Peace River span. On October 16, 1957, one anchorage began to slide, and within hours the center span dropped into the river. Nothing could be salvaged, and a steel truss bridge replaced it in 1960. Furthermore, the Peace River Bridge actually had been completed before workers finished salvaging the Tacoma Narrows wire (started November 7, 1942; ended March 1943), and the Peace River cables were made of pre-fabricated, helical strand wire, quite different from the parallel strand wire of the Tacoma Bridge. Nonetheless, the Peace River collapse apparently inspired the myth that somehow the two ill-fated suspension bridges shared a common destiny, supposedly linked by salvaged parts from Galloping Gertie.[10]

ENGINEERING CHALLENGES FOR GERTIE'S REPLACEMENT

Engineers faced major challenges in building the second Narrows Bridge. First, they had to understand what happened to the 1940 bridge and adopt a structurally sound design that would not meet the same fate. Second, they had to find out whether remnants of the old bridge, especially the piers, could be reused. Third, the cost of the new Narrows Bridge would have to be paid for by income it generated in tolls.

In July 1941, the WTBA appointed Charles E. Andrew as the principal engineer and chairman of the consulting board. Other board members included Dr. Theodore von

Karman, Glenn Woodruff, and the firm of Sverdrup & Parcel of Chicago. Andrew picked Dexter R. Smith as the chief design engineer for the new structure. By October, the State had a preliminary draft of a new design, which roughly resembled Clark Eldridge's original effort. The proposed replacement bridge with a deep, open, stiffening truss would cost about $7 million.

The U.S. Navy lobbied for a combination highway and railroad bridge. A steel cantilever-type structure appealed to the Navy's needs more than a suspension span did, but this would require twice the amount of steel and a year longer to build, adding as much as $8.5 million to the cost. It was not an attractive solution to the Washington Toll Bridge Authority and never gained serious support.

The proposed design for the new Narrows Bridge needed testing. Because the engineers knew so little about the forces that had destroyed Galloping Gertie, a purely mathematical solution was not possible. "The only way to attack the problem," noted Charles Andrew, "was to first design a bridge, then build a model of that design and actually subject it to wind action in a specially built wind tunnel." If the tests indicated problems, then the design could be changed and the model retested until a good design emerged.[11]

The testing fell primarily to Professor Burt Farquharson at the University of Washington, while other tests were conducted at the California Institute of Technology in Pasadena by aerodynamicist Theodore von Karman and his assistant, Louis Dunn. Dexter Smith and the State's bridge design team consulted extensively with Farquharson and his research group at the university. Their pioneering work was on the cutting edge of modern bridge aerodynamics.

Beginning in late 1941, throughout World War II, and occasionally afterward to 1950, Farquharson studied the 1940 span and the new proposed Narrows Bridge. By 1943, he was working in the specially built Structural Research Laboratory on the university campus. The building housed a wind tunnel and a 100-foot-long scale model of the new bridge, plus several section models. First, Farquharson confirmed that the 1940 bridge had collapsed because of its excessive flexibility and susceptibility to aerodynamic forces, laying the foundation for continued research. If a dynamic scale model of the proposed bridge design could pass Farquharson's wind tunnel testing, then Smith and his bridge engineers could erect the real bridge with confidence at the Tacoma Narrows.

They envisioned a new span designed to offer the least wind resistance. The solution would be to use deep, open, stiffening trusses with trussed floor beams. The truss members would be shallow, to avoid creating any large, solid surfaces such as the ones associated with the wind-caused failure of the 1940 bridge.

After Farquharson built the 1:50 full-scale model plus three sectional models of Smith's design, testing began in November 1943 and continued through 1945, with occasional experiments conducted as late as 1950. Farquharson's studies included subjecting the models, in approximately 200 different configurations, to wind forces striking the bridge at angles up to plus-and-minus 45 degrees perpendicular to the deck. This wide range of wind angles helped give the new bridge design even greater stability. (Today, design engineers typically use a narrower range of plus-to-minus 5 degrees.) The research proved that the proposed bridge would be far more stable than Galloping Gertie. Farquharson then tested the model fitted with strips of open wind grating to permit freer airflow and minimize the wind's effects. It worked—the model now showed virtually no torsional movement.

Smith and Farquharson decided to take additional steps to eliminate as much vertical and twisting motion in the model as possible. First, they added a bottom lateral bracing system to the stiffening truss to increase torsional stiffness. Second, they added hydraulic shock absorbers at three strategic points in the structure: at mid-span, between the main span and side spans, and at each tower.

The Structural Research Laboratory at the University of Washington, where extensive wind tunnel testing was conducted on models of the replacement bridge design.
Washington State Archives

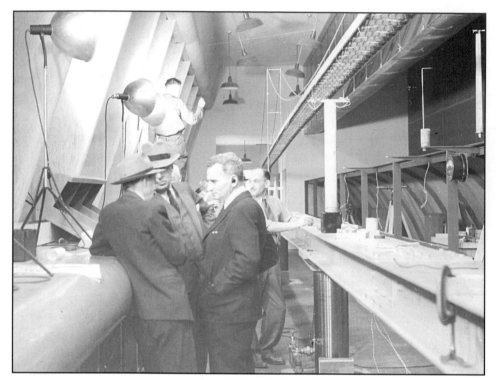

Engineers confer in the wind tunnel, with the full-scale model of the replacement bridge at right, ca. 1943. Left to right: unidentified, Charles Andrew, Theodore von Karman, and F. Burt Farquharson.

UW Libraries, Special Collections, FAR043

These mechanical devices would enhance the design's natural damping ability. The tests from 1941 to 1947 cost more than $88,000, but the extensive investigation gave the engineers confidence in the new design. The proposed bridge would stand safe and solid in winds up to 127 mph in 3-second gusts.

The existing piers posed the second engineering challenge. Would they support the proposed four-lane bridge, which was 60 percent heavier than the old superstructure? The piers had been stressed by the collapse of the 1940 bridge, as well as by 17 earthquakes that struck the area between 1939 and 1946, two of which reached 5 on the Richter scale. Engineers discovered that the piers would prove to be solid foundations for the heavier new span.[12]

A DECADE LATER

Almost a decade would elapse between the first bridge's collapse in November 1940 and the completion of the replacement span in October 1950. Insurance litigation was only the first of a series of events that slowed work.

World War II also held up progress and the salvage efforts did not end until June 1943.

Although the State had a preliminary draft design for the new bridge ready by October 1941, its testing was delayed. The Japanese attack on Pearl Harbor in December and Professor Farquharson's involvement in war related projects hindered tests. Financing the second Narrows Bridge also proved difficult. After the end of World War II, many delayed major construction projects competed for bond financing. The needs of Tacoma, Bremerton, and other area residents were not high on financiers' priority lists.

Under the careful guidance of Charles Andrew and Dexter Smith, the main design plans were completed on December 5, 1945, and by April 1946 the WTBA approved revised designs, with a projected cost of $8.5 million. Some minor changes to the plans continued until September 1946. Steel was in short supply in the immediate post-World War II years and the State had difficulty arranging for insurance.

On April 30, 1947, Governor Mon Wallgren announced that insurance for the bridge finally had been arranged, with 100 companies participating. Final designs were prepared, but not until August 1947 was the State able to request bids for the new bridge. By this time the cost had gone up. The price tag for construction—$11.2 million—was one-third more than what the Washington Toll Bridge Authority earlier had estimated. On October 15, the WTBA opened bids for construction. Low bidders were the Bethlehem Pacific Coast Steel Corporation ($8,263,904 to build the superstructure) and John A. Roebling's Sons Company of San Francisco ($2,932,681 for the cable construction). However, final financing had not yet been arranged and again the start of construction was delayed.

The tide turned when Pierce County contributed $1.5 million to a bond guarantee fund. The final construction cost estimate, completed just prior to the bond issue, reached $13,738,000. In December 1947, the WTBA offered a bond issue of $14 million, with revenues from tolls to be used to pay for the bonds. Finally, on March 12, 1948, State officials completed the bond financing.

Now, too, steel was more readily available. At long last, events began to move quickly. On March 31 and April 1, the WTBA awarded the contracts and on April 9, 1948, bulldozers rumbled over the ground at the east anchorage and began clearing the site and moving earth to start construction of the replacement bridge.

Notes

1. "Row over Bridge's Collapse," *Tacoma News Tribune*, November 11, 1940; "US Loaning Agents Held Insistent on Own Design," *Seattle Times*, November 8, 1940.

2. "Conversation between T.L. Stokes of the Scripps-Howard Syndicate and Harry M. Brown, December 10, 1940," Project Files, Tacoma Narrows Bridge, 162.2.3, General Records of the Federal Works Agency, National Archives and Records Administration, Washington, D.C. Docket 1807-F for the Tacoma Narrows Bridge Project was started when the span was first proposed and it remained the PWA's principal file for the bridge. It initially contained the original application, as well as the reports of PWA inspector D.L. Glenn, change orders, related correspondence, and other documents. Approximately three-quarters of the records appear to have been accidentally destroyed in 1943. Today, the Project Files for the Tacoma Narrows Bridge contain 408 pages of correspondence, dating between ca. November 1940 and December 1941, and 200 items in a "Clipping File"; letter, Gene Morris (NARA Civilian Records) to the author, June 8, 2005.

3. "Questions on Tacoma Bridge," Project Files, Tacoma Narrows Bridge, 162.2.3, General Records of the Federal Works Agency, National Archives and Records Administration, Washington, D.C.

4. "Government Warned about Narrows Span by Own Engineer, Is Disclosure," *Tacoma News Tribune*, January 11, 1941. Some of James Roper's reports and Clark Eldridge's correspondence are in "Tacoma Narrows Bridge, Progress Reports 1940," Box 102, von Karman Papers, Caltech; Advisory Board, *Failure of the Tacoma Narrows Bridge*, Section B, 19. Unfortunately, Glenn's reports seem to have disappeared.

5. Letter, Jas. A. Davis (WTBA) to Leon Moisseiff, February 21, 1941, Box 53, WTBA, WSA; J.J. Madigan's letter is in Advisory Board, *Failure of the Tacoma Narrows Bridge*, Appendix III.

6. "Row over Bridge's Collapse," *Tacoma News Tribune*, November 11, 1940. Four decades later, Clark Eldridge repeated this observation regarding Mosseiff and Moran & Proctor in his "Autobiography."

7. Charles Andrew, "Redesign of Tacoma Narrows Bridge," *Engineering News-Record* 135 (November 29, 1945): 716–21.

8. "Shuttle Ferry Service Replaces Fallen Bridge," *Tacoma News Tribune*, November 8, 1940.

9. Charles Andrew, "General Report on the Design of the Tacoma Narrows Bridge," January 15, 1942, Box 43, WTBA, WSA; Gunns, 165, 168–69; "Fraud Charged over Narrows Span Insurance," *Seattle Post-Intelligencer*, December 4, 1940; "Narrows Span Total Loss, Board Claims," *Seattle Post-Intelligencer*, March 11, 1941; "Hopes for Rebuilding Narrows Span Fade," *Seattle Post-Intelligencer*, February 3, 1942; "Narrows Bridge Plan Kept Alive," *Seattle Times*, February 3, 1942; "State to Sell Steel of Tacoma Bridge," *Seattle Post-Intelligencer*, May 9, 1942; "Narrows Bridge to Aid Drive," *Seattle Post-Intelligencer*, October 10, 1942.

10. Scott, *In the Wake of Tacoma*, 95–97; Arnie Colby, interview, September 2005; e-mail, Michael Cegelis, American Bridge Company, to the author, September 19, 2005; inter-departmental correspondence, Tom Beell to Bill D., November 4, 1965, Box B6, WSDOT Library records, WSA.

11. Letter, Charles Andrew to Jas. A. Davis, August 3, 1945, Box 42, WTBA, WSA.

12. Charles Andrew, *Final Report on Tacoma Narrows Bridge* (Tacoma, Washington: Washington Toll Bridge Authority, 1952); Farquharson, *Aerodynamic Stability*; F. Burt Farquharson, "Lessons in Bridge Design Taught by Aerodynamic Studies," *Civil Engineering* (August 1946): 344–45; "Bridge Wind Experiments on Tacoma Span to Begin," *Seattle Star*, April 30, 1945; "Model to Show Why Narrows Bridge Fell," *Seattle Post-Intelligencer*, September 17, 1942; "'Gale' Sways Bridge Model," *Seattle Times*, October 4, 1942; "Supplemental Tests on the Dynamic Model of the Original Tacoma Narrows Bridge," Seattle, University of Washington, Structural Research Laboratory [1943]; "Tacoma Bridge Oscillations Being Studied by Model," *Engineering News-Record* 126 (April 24, 1941): 139; "Tacoma Narrows Bridge: Reconstruction to Follow Design Resulting from Extensive Wind Tunnel Research," *Roads and Streets* 90 (December 1947): 88–90.

The "epoch-making" 1950 Narrows Bridge.
WSDOT

CHAPTER TEN

RISE OF A NEW BRIDGE

AN "EPOCH-MAKING" NEW SPAN

Thanks to lessons learned from the failure of its predecessor, the 1950 Narrows Bridge significantly affected the course of suspension span design and engineering. The years of aerodynamics research, and the new mathematical knowledge of vibrations and wave phenomena, ushered in an era of more stable spans. The next generation of large suspension bridges featured deep and rigid stiffening trusses. Completion of the 1950 Narrows Bridge soon was followed by the Delaware Memorial Bridge in 1951, the Chesapeake Bay Bridge in 1952 (since 1957, named the William Preston Lane Jr. Memorial Bridge), and David Steinman's great Mackinac Strait Bridge completed in 1957.

"Epoch-making" is how Richard Scott, author of the landmark study *In the Wake of Tacoma: Suspension Bridges and the Quest for Aerodynamic Stability* (2001), describes the 1950 span. It represented a remarkable achievement with unprecedented design, engineering, function, and stability. Its aesthetic appeal marked a milestone as well. The visually stunning "Sturdy Gertie" helped shift popular perceptions about "beautiful" suspension spans.

Overall, the 1950 Narrows Bridge has a suspended structure of 5,000 feet—a center span of 2,800 feet and two side spans of 1,100 feet (the same as its predecessor, since the same piers were used). The motion damping devices tested in Farquharson's wind tunnel all appeared in the finished structure. The new four-lane bridge had several features that made it 58 times more rigid than the 1940 bridge, immediately earning the span a distinctive place in engineering history. In fact, in 1950 the Tacoma Narrows Bridge was the

BRIDGE CENTER SPAN RATIOS

Bridge	Width to length (of center span)	Girder depth to length (of center span)
1940 Narrows Bridge	1:72	1:350
1950 Narrows Bridge	1:46	1:85
Golden Gate Bridge	1:47	1:168
George Washington Bridge	1:33	1:120[a]
Bronx-Whitestone Bridge	1:31	1:209[b]

a. After second deck was added.

b. Before truss was added above plate girder.

Elevation detail from Dexter Smith's original layout and plan, December 5, 1945. *WSDOT*

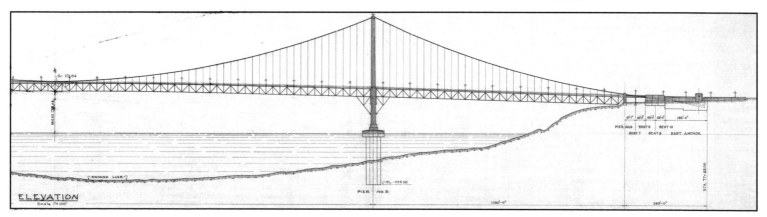

A boy stands in a giant eight-foot-diameter reel of spinning wire, October 1949.
WSDOT

most technically advanced long span suspension bridge in the world.[1]

INNOVATIONS AND SPECIAL FEATURES

- The prominent 33-foot-deep steel Warren stiffening trusses gave a depth-to-center span ratio of 1 to 85, the deepest stiffening system on a major suspension span since the 1909 Manhattan Bridge.
- Double (top and bottom) lateral bracing of the stiffening trusses, combined with the 33-foot-deep stiffening trusses, gave the bridge exceptional torsional rigidity.
- Wind grates 33-inches wide separated all four traffic lanes on the deck (i.e., three slots of open steel grating), with strips 19-inches wide along each curb.
- Hydraulic shock absorbers were placed at three strategic points in the structure: (1) at mid-span (at the main cable center tie between the main suspension cables and the top of the stiffening truss; with six devices per cable, it was a first for a long suspension bridge), (2) between the top chords of the main span and side span stiffening trusses, and (3) at the towers, where they join the bottom of the deck truss.

East tower (#5) nears completion, April 11, 1949, two days before an earthquake struck.
WSDOT

- The ends of the west and east spans were anchored securely to solid ground.
- A cable sag ratio of 1:12 required the towers to be higher than for the 1940 bridge, which had a sag ratio of 1:10. The greater sag ratio lessened stress on the main cables and therefore reduced the required amount of cable wire and anchorage concrete.

BUILDING THE SECOND NARROWS BRIDGE

Like a phoenix, the second bridge at the Narrows began to rise on April 9, 1948, when construction started with site preparations at the east anchorage. Some 30 months later on October 14, 1950, the span would open to the public. Once again, the Narrows would be the site of the third longest suspension bridge in the world.

Piers

The piers (originally designed by Clark Eldridge for Galloping Gertie) supported the new span's towers. The twin steel legs of each

tower stood 60-feet apart center-to-center. The tops of the original concrete pedestals, however, were too close to the water; in Gertie's short life, the tower legs had shown signs of salt-water corrosion. The new pedestals were the same width, but taller by 17 feet to keep the tower legs free of salt spray.

The towers rose 58 feet higher than those of the 1940 bridge. The additional weight of the superstructure imposed a load 1.6 times greater than Galloping Gertie's load on the original piers. This was actually better, because the weight was more evenly distributed by the greater width between the tower legs. Also, the new design increased the dead load pressure of the piers on the underlying bed of the Narrows by only 6 percent.

Towers

The piers were finished and ready for Bethlehem Steel to start erecting the towers by the end of December 1948. A "crawler crane,"

Setting the tower base plate, 1948. At left is State engineer Harry Cornelius; second from right, ironworker Slim Bonner.

also called a "creeper" or "traveler derrick," advanced upward along the tower legs as the crews erected them. Each tower consisted of hollow steel cells stacked one on top of another. The four columns of the cell sections were arranged to form a hollow core in the center, with each section measuring 40-feet long and weighing 27 tons. Viewed in cross-section, the four cells in each leg formed a cross shape. To stabilize the towers during construction, temporary "outriggers" were added until the cables and deck trusses could be completed.

West tower (#4) begins to rise off its pier; east tower (#5) in the background, April 22, 1949. *WSDOT*

When the top cross-bracing was completed, a Chicago boom hoisted the 28-ton cast-steel cable saddles and placed them at the top of each leg. The saddles were secured by 36 bolts.

The large "X"-shaped cross braces of the towers varied in size; below the truss, all three were 45-feet high; above the roadway, the ones that motorists would see the most, were progressively smaller. The first measured 26 feet 6 inches high, the middle one, 24 feet 6 inches, and the one at the top of the tower, 23 feet 5 inches. Tower erection began January 1, 1949, and was completed by late June.

Anchorages

The new bridge had a much larger cable load, increased from the original 28 million pounds to 36 million pounds, which required modification of the 1940 anchorages. The old anchorages (originally with cables spaced 39-feet apart), were retrofitted for the new span's 60-foot spacing between the cables. The original anchorages became the core of new heavier and wider 54,000-ton anchor blocks. The anchorages included 62-foot-long eye-bars fitted with 26-inch-diameter shoes embedded into new concrete.

East anchorage under construction, showing final concrete pour to encase the 1940 bridge's anchorage, April 4, 1949.
WSDOT

Cable spinning starts,
October 1949.
WSDOT

Catwalks from the east tower (#5) to the west tower (#4), October 28, 1949.
WSDOT

Cables

Once the towers and cable saddles were in place, spinning began for the pair of 20¼-inch diameter main suspension cables, spaced 60-feet apart across the bridge. Each cable would contain 19 strands of 458 No. 6 gauge wires. The strands were looped around the eye-bar shoes to run continuously from anchorage to anchorage.

Cable construction officially began on July 18, 1949, as preparations started immediately on two fronts: (1) building the catwalks for the crewmen during cable spinning; (2) erecting the spinning towers, tramways, and related facilities at the east anchorage so that wire would unwind at the correct speed and tension. The 10-foot-wide catwalks—consisting of one-inch diameter wire cables, cyclone wire fencing, and a four-foot-wide center section of wooden slats—were pre-fabricated in 200-foot sections.

In order to supply the 100-million feet of galvanized wire needed to spin the main cables, Roebling's Sons Company set up a temporary reeling plant on tide flats near the Tacoma docks. Large 350-lb. coils of wire arrived by rail from Trenton, New Jersey.

Nineteen strands of the south suspension cable splay to the anchor bars, after spinning and compacting. The tram system that had carried the spinning wheel now hauls a box of materials to the tower, March 1950.
WSDOT

Main cable compacting, February 1950.
WSDOT

Art Knoll (left) and Harry Takahara (right) attaching cable bands to the north main suspension cable, March 1950.
WSDOT

Workers transferred the coils to intermediate reels, then wound the wire onto the final reels to hold it at a uniform tension. The giant final reels, 8 feet in diameter, each held 36 miles of wire and weighed 9 tons. The cable spinning crew spun the first cable wire in October 1949, and completed the job on January 16, 1950, in the midst of a harsh winter.

Suspended Structure

The 33-foot-deep Warren stiffening truss system was assembled at the site from shop-fabricated components. Four rolling derricks (two per tower) moved in opposite directions from each tower—i.e., two traveler operations and riveting crews worked from each tower pier toward

WTBA Chief Inspector Ken Arkin (next to car) observing the bridge from the west end. The suspender cables are hung and ready for the suspended structure to be installed, March 14, 1950.
WSDOT

the center of the main span, while two other crews worked from the towers toward the east and west shorelines. First, workmen placed the top and bottom chords and their diagonal bracing. Next, the floor beams were fitted between the chords, and deck stringers were laid lengthwise on top of the beams. Finally, crews pinned the members in place and the riveting gangs finished the process. Workers met at the mid-point in the center span to close the deck at midnight on June 1, 1950.

The road deck measured 46 feet 9 inches (curb-to-curb) for four traffic lanes, plus two sidewalks, each 3-feet 10-inches wide. Steel reinforcing rods were placed in the roadway before workmen completed the deck with a lightweight concrete slab 5¾-inches thick,

topped by a ⅝-inch asphalt riding surface. Open, steel wind grates were installed between driving lanes and at the curbs.

Suspender Cables and Cable Bands

As each section of the stiffening truss was assembled, crews hung the suspender cables. On either side of the bridge, each set of cables (in pairs) were looped over the main cable, giving the appearance, when in place, of being four cables. These pairs of 1⅜-inch diameter cables were attached to the stiffening truss at 32-foot intervals by zinc connectors called "jewels." The placing of the cable bands for each two-cable set was completed by March 7, 1950.

Motion Damping Devices

Workmen also installed mechanical motion damping devices developed and tested by bridge engineers and Professor Farquharson. These innovative hydraulic shock absorbers were placed at strategic points in the structure.

Sand Blasting and Painting

The bridge was painted a color called "Narrows Green." The sub-contractor, H.P. Fisher & Sons Company of Seattle, painted the suspender cables, cable bands, and various other steel parts, applying three coats.

Cross section and elevation of suspended structure, December 5, 1945.
WSDOT

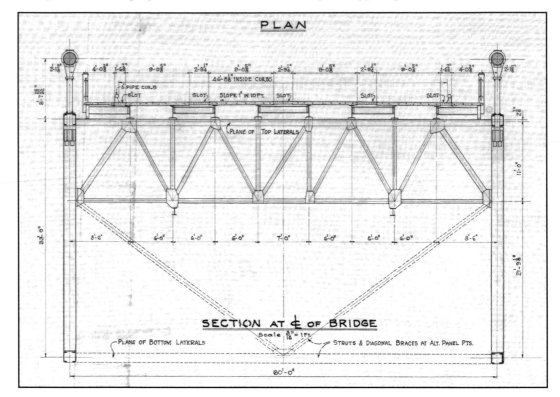

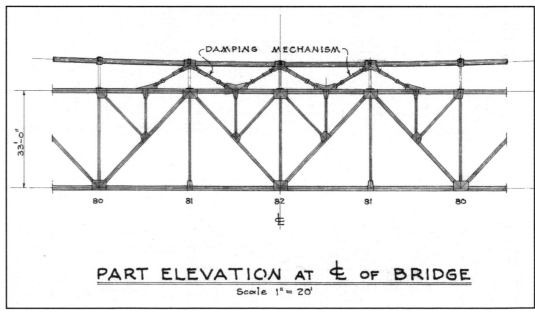

Earthquake, April 13, 1949

In addition to the imposing engineering challenges, the builders faced other difficulties that slowed their progress. First was an earthquake. The Puget Sound region is prone to earthquakes, though typically they are localized and of relatively low intensity. From 1903 to 1949, instruments at the University of Washington recorded about 85 tremblers. Only 4 had been strong enough to crack brick or masonry buildings. Suspension spans, of course, are designed to be flexible. The Narrows Bridge was built to withstand earthquakes more than twice as strong as ones that would flatten an average brick building.

By April 13, 1949, work had progressed well. The piers were done and the anchorages neared completion. About 60 workmen were engaged in finishing the final sections of the east tower (#5). Only several steps remained— placement of the top strut between the tower legs and pinning and bolting a second strut above the roadway (prior to riveting), riveting the tower joints in the top 150 feet of the tower, and placing and securing the cable saddles.

Meanwhile, the north cable saddle had been put up on the tower, ahead of schedule. The contractor hoped to save two days of work time by placing the 28-ton cast-steel cable saddle on the tower top, against the advice of engineers. Normally, a saddle would be bolted down soon after being lifted into place. In this case, however, the cable saddle was blocked up on timbers some 30-inches high to allow space for workers to rivet the slab steel plate under the saddle's location. Only four temporary bolts attached the saddle to the timbers. At this point in its erection, the east tower was "free standing"—i.e., without lateral support, making it insecure and vulnerable.

That morning, an earthquake measuring 7.1 on the Richter scale shook the Puget Sound region. In Tacoma, buildings swayed, while bricks and debris from weaker structures fell on cars and into the streets. An

A boom moves a truss section into place during the erection of the suspended structure, April 1950.
Courtesy of Earl White

11-year-old boy was killed by a falling brick as he fled Lowell Elementary School. At the Narrows, the vibration rocked the new east tower, swaying it six feet from perpendicular. Ironworker Phil Orlando was standing on top of the north cable saddle, giving directions to the Chicago boom operator below. When the trembler started, the saddle slid across its flat platform, shearing the bolts. Orlando scrambled across the platform to escape. The saddle slid back, and Orlando again ran back across. A third time the saddle moved across its base, and Orlando got off just ahead of it.

The cable saddle plunged off the tower, fell 500 feet, bored through one end of a barge, and sank 135 feet into the Sound, coming to rest about 75 feet from the east pier. The barge foundered, taking with it a compressor and numerous tools. Orlando figured he had narrowly escaped death three times. He climbed down the ladder, and told his boss he was quitting and they could mail his paycheck. Nobody blamed him—other equally stalwart bridgemen had the jitters for a few days.

Before long, with no new aftershocks, the crews regained confidence and work returned to normal. Finding the cable saddle proved no easy feat. Divers could work only during slack tides, which came at night, so it took three full days and four trips to the channel bottom to locate the cable saddle. They attached a line and the errant mass of steel was hoisted to the surface. It had suffered a small bend on one corner of the base, which could be repaired.

After 10 days, workers placed the cable saddle back on the tower. By early June, both saddles were ready to be bolted securely in place on top of the east tower.[2]

FIRE AND ICE

In the first week of June, work on the east tower was 98 percent complete, with rivet gangs driving some 71,300 rivets. The Chicago boom had moved to roadway level in preparation for starting the truss assembly. More than 70 men were rapidly erecting the west tower (#4), already counting more than 10,900 rivets in the structure. Then, in the early morning hours of Wednesday, June 8, 1949, the creosote-soaked timber fenders at the base of the west tower caught fire. Flames

View from the east tower of the suspended structure work, May 1950.
Courtesy of Earl White

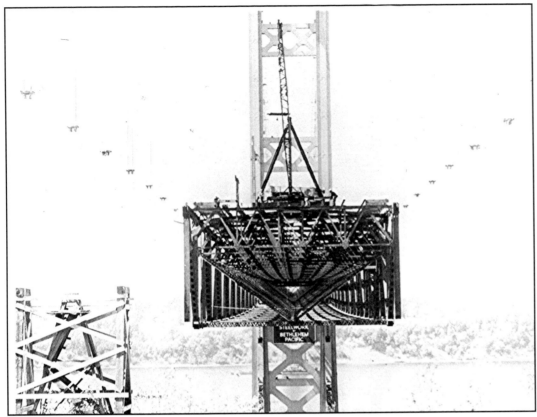

Building the suspended structure—looking west from the east anchorage toward the east tower, April 1950.
Courtesy of Earl White

rose more than a hundred feet, and the heat ignited wood scaffolding some 400 feet above.

The spectacular blaze, probably started by a faulty electrical system or a transformer explosion, caused $200,000 in damage. At first it was feared that the fire had damaged the tower steel, which would mean completely dismantling it. But the heat only had a minor effect on a couple of the steel plates, which were easily repaired. Most of the damage was to the fender and some of the contractor's equipment—the Chicago boom, compressors, motors, and air tools. In all, the blaze disrupted bridgework for two weeks.

A half-year later, severe winter weather also caused delays. Engineers already had concerns about starting the cable spinning in the coldest months of the year. The winter of 1949–1950, one of the harshest ever recorded in the area, proved their worries valid. For six weeks, rain and snow driven by high winds lashed through the Narrows. In mid-January, particularly, freezing northerly gusts up to 70

mph pounded the construction site. At one point, men had to remove inch-thick ice from steel plates and beams so they could do their jobs.

It was especially rough for the cable spinning gangs in late January. As Arnie Colby, a survey crewmember, recalls, "That winter was terrible. There was rain and freezing rain. Ice went in your face and down your sleeves. It was miserable, miserable."

Warren truss of the 1950 bridge.
WSDOT

East anchorage elevation, December 5, 1945. *WSDOT*

At times, too much ice lodged in the strands, making it impossible to compact the cables. In other instances, workers with blow-torches were able to thaw out cable strands for adjustment and banding. Ice made the 10-foot-wide catwalks dangerous. Earl White recalls, "You'd see guys go walk down it and their feet slide out from under them, and they'd slide 15 to 20 feet before they could grab something to stop themselves." By February the cable spinning contractor, John A. Roebling's Sons Company, was fighting a two-month delay.[3]

Four Workers Died

When men lost their lives during the building of the 1950 Narrows Bridge, their fellow workers honored them with the traditional gesture of respect—crewmembers quit work for the day and went downtown to hold a wake. Workers faced the greatest danger during the stiffening truss construction—two of the four men died during this phase in 1950.[4]

Robert E. Drake, May 24, 1948. First to lose his life was carpenter Robert Drake, employed by the Woodworth Company. He and fellow workers had been busy at the west anchorage. On May 24, Drake happened to be in the wrong place at the worst possible moment. He was standing below a derrick just as a cable broke, sending the boom crashing down on him.

Lawrence S. "Stuart" Gale, April 7, 1950. It was a Friday, about 3:45 p.m. Stuart Gale was helping to connect stiffening truss sections. Without warning, a temporary weld on the lower chord cross strut at panel 33 broke, sending Gale to his death. A young ironworker, Harold Peterson, jumped into the water and swam furiously toward Gale, but it was futile. In the swift current, Gale's body floated away too fast, and soon, his heavy clothes and tools pulled him below the surface.

Fellow ironworker Earl White was there and remembered the tragedy. "Gale died when they were working on the sections down below deck, where the diagonal and bottom chords were hooked together. These sections had temporary welds on them that were put on in the shops. They had swung this one section in, and Stuart Gale looked, and he called up to the young foreman, Danny Lowe. He said, 'These welds don't look very good down here.' It was Gale's first day on this part of the job, connecting the roadbed. Danny said, 'Well, Gale, they've been holding all the way across. We haven't had no problem.' A second time

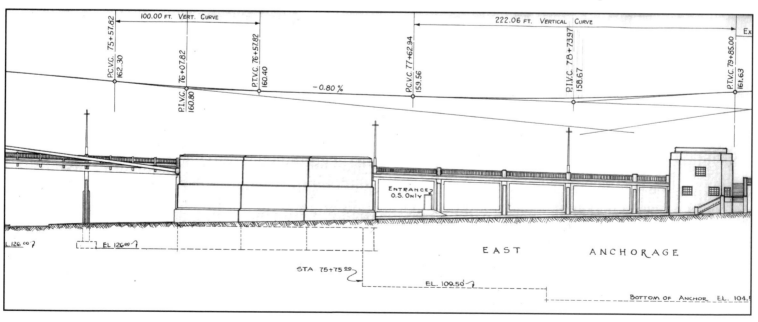

Gale hollered up. And, Danny said, 'Cut 'er loose.' And, when Gale did, well, 40 ton of iron and him went in the hole. You see, safety nets wouldn't have stopped him…they wouldn't have caught Gale and 40 ton of steel. He'd have gone right through. Danny was really hurt over this. He could never live that down."

Stuart Gale was 36 years old. A memorial service was held on a boat floating beneath the unfinished stiffening truss. Gale's wife and 3-year-old daughter sat quietly. At the end of the ceremony, Mrs. Gale rose to her feet and cast a flower wreath into the swirling Narrows.

Glen "Whitey" Davis, June 6, 1950. Another ironworker who helped build the suspended structure was Glen "Whitey" Davis. Earl White worked on the same crew as Davis and also grimly recalls his death. "He and I were real close friends. We had finished the deck and were starting to load timbers for the deck that the cement trucks would drive on. I had swung a big load of timbers in to him.

When he stepped, he missed and went all the way down. God, when he hit, it sounded like an artillery piece went off."

Ray Bradley, July 31, 1950. The Bethlehem Pacific Coast Steel's crew was down to about 44 workers on an average day in late July. Some men were putting in roadway grating, others placed the expansion joints, and some were installing brackets for the sidewalk on the west approach. On the last day of July, heavy rains soaked the workers. Ray Bradley, a welder in his mid-fifties, was working on an expansion joint. He reached down, grabbed the lead Direct Current wire for a welding machine, stuck it under the armpit of his soggy jacket, and started dragging the line along the bridge. Suddenly he slumped over. Bradley died on the way to the hospital. According to the Pierce County Deputy Coroner, John Wolf, he had suffered a fatal heart attack, but the bridgemen believed he was electrocuted.

Notes

1. The features and construction process of the 1950 Narrows Bridge are described in: Andrew, *Final Report on Tacoma Narrows Bridge,* 1952; William Addis, "Design Revolutions in the History of Tension Structures," *Structural Engineering Review* 6 (February 1994): 1–10; O.H. Ammann, Charles A. Ellis, and F.H. Frankland, "Unusual Design Problems—Second Tacoma Narrows Bridge," *Proceedings of the American Society for Civil Engineers* 74 (June 1948): 985–92; Charles Andrew, "Design of a Suspension Structure to Replace the Former Narrows Bridge—Part 1," *Pacific Builder and Engineer* 51 (October 1945): 43–45; Charles Andrew, "Redesign of Tacoma Narrows Bridge," *Engineering News-Record* 135 (November 29, 1945): 716–21; Charles Andrew, "Tacoma Narrows Bridge Number II…The Nation's First Suspension Bridge Designed to be Aerodynamically Stable," *Pacific Builder and Engineer* (October 1950): 54–57, 101; Charles Andrew, "Unusual Design Problems—Second Tacoma Narrows Bridge," *Proceedings of the American Society of Civil Engineers* 73 (December 1947): 1483–97; "Cable Spinning at Tacoma Narrows," *Engineering News-Record* 144 (February 16, 1950): 44–45; "Construction Features of the Tacoma Narrows Bridge," *Pacific Builder and Engineer* 52 (January 1946): 44–49; "High-Strength, Lightweight Deck for New Tacoma Narrows Bridge," *Engineering News-Record* 146 (January 11, 1951): 34; Harold W. Hills, "The Techniques of Cable Spinning as Exemplified at the New Tacoma Narrows Bridge," *Western Construction* 57 (February 1951): 78–81; Edward Horwood, "Cable Spinning Operations Underway at Tacoma Narrows Bridge," *Pacific Builder and Engineer* 55 (November 1949): 44–47; "'Jinx' Bridge Going Up Again," *Western Construction News* 24 (August 15, 1949): 61–63; A.R. MacPhersan, "Construction Begins on New Tacoma Narrows Bridge," *Roads and Streets* 92 (January 1949): 63–65; Douglas B. Mauldin, "'Galloping Gertie's' Legacy," *The Highway User* (April 1965): 21–23; Elmer C. Vogel, "New Tacoma Narrows Bridge Called Forerunner of New, Safe Type of Suspension Structure," *Seattle Times,* October 8, 1950; "Narrows Bridge Ready Sept. 1950," *Seattle Times,* March 13, 1948; "Aerodynamic Design Makes New Narrows Span Safe," *Tacoma News Tribune,* February 14, 1950; "Progress of Narrows Bridge Shown," *Tacoma News Tribune,* February 14, 1950; "Narrows Bridge Edition," *Tacoma News Tribune,* October 13, 1950; see also "Aerodynamic Stability" folders, Box 38, "Redesign" folders, Box 57, and "Testing, University of Washington, 1940–44" folders, Box 60, WTBA, WSA.

2. Progress Reports, April 18, 20, 25, and May 7, 1949, Box 56, WTBA, WSA; Earl White, interviews, October 2003, November 2004, May 2004; Gotchy, *Bridging the Narrows,* 64–65; "Falls 180 Feet into Sound," *Tacoma News Tribune,* April 10, 1950; Jackson Durkee, telephone interview, November 8, 2005.

3. "Steel Tower at Tacoma Narrows Bridge Undamaged by $200,000 Blaze," *Pacific Builder and Engineer* 55 (July 1949): 68; Earl White, interviews, October 2003, November 2004, May 2005, September 2005; Earl White quoted in John C. Ryan, "Narrows' Runaway Catwalk Pales Next to '49's Big Splash," *Daily Journal of Commerce,* August 18, 2005.

4. Earl White, interviews, October 2003, November 2004, May 2005, September 2005; Gotchy, *Bridging the Narrows,* 64–65; "Killed on Bridge Job," *Tacoma News Tribune,* May 26, 1948; "Man Killed at Narrows Span, Falls 180 Feet into Sound," *Tacoma News Tribune,* April 10, 1950; [Glen Davis notice] *Tacoma News Tribune,* June 7, 1950; Progress Reports, April 11, June 12, and August 12, 1950, Box 56, WTBA, WSA; "Heart Attack Claims Worker on Narrows," *Tacoma News Tribune,* August 1, 1950.

STATISTICAL PROFILE OF THE 1950 TACOMA NARROWS BRIDGE

General

Cost	$14,011,384
Total structure length	5,979 feet
Suspension section	5,000 feet
Center span	2,800 feet
Shore suspension spans (2), each	1,100 feet
East approach and anchorage	365 feet
West approach and anchorage	614 feet

Suspended Structure

Roadway height above water (mean sea level)	200 feet at towers, 220 feet at center span
Center span vertical clearance above water (mean sea level)	184 feet 6 inches
Weight of center span	7,250 lb./ft
Traffic lanes	4
Width between cables	60 feet
Width of sidewalks (2), each	3 feet 10 inches
Width of roadway	49 feet 10 inches
Thickness of roadway	6⅜ inches, reinforced concrete
Suspender cables, intervals	32 feet
Depth of stiffening girder	33 feet
Number of girders and type	2 Warren trusses
Ratio, deck width to center span	1:46
Ratio, deck depth to center span	1:85

Anchorages

Weight of each anchorage	66,000 tons
Concrete in each anchorage	25,000 cu. yds.
Structural steel in both anchorages	901 tons
West anchorage (concrete anchor block and gallery)	164 feet long
East anchorage (concrete anchor block and gallery), approach, administration buildings, and toll house	185 feet long
West anchorage, construction and cost	Woodworth & Co., $406,000 (est.)
East anchorage, construction and cost	Woodworth & Co., $386,000 (est.)
Cable anchor bars	38 in each anchorage, 62 feet long

Cables

Diameter of main suspension cable	20¼ inches
Weight of main suspension cable	5,441 tons
Weight sustained by cables	18,160 tons
Sag ratio (vertical distance between tower top and cable elevation at mid-span, as a ratio of main span length)	1:10
Number of wire strands in each cable	19
Number of no. 6 wires in each strand	458
Number of no. 6 wires each cable	8,705
Total length of wire	104,094,390 feet (19,715 miles)
Cost of cables, bands, suspenders, fittings, etc.	$2,732,773

Towers

Height above water (mean sea level)	507 feet
Height above piers	467 feet
Height above roadway	307 feet
Weight of each tower	2,675 tons
Legs (2 per tower), each	14 feet 8 inches by 14 feet 8 inches
Cost of both towers	$1,977,167

Piers

West Pier (#4), total height	215 feet
West Pier (#4), depth of water	120 feet
West Pier (#4), penetration at bottom	55 feet
East Pier (#5), total height	265 feet
East Pier (#5), depth of water	135 feet
East Pier (#5), penetration at bottom	90 feet
Area of pier at top	118 feet 11 inches by 65 feet 11 inches

Materials

Structural steel	17,000 tons
Cable wire	5,241 tons
Concrete in both anchorages	50,000 cu. yds.

Night work during cable spinning, February 1950.
WSDOT

CHAPTER ELEVEN

PEOPLE OF THE COLLAPSE AND AFTERMATH, 1940–1950

HOWARD CLIFFORD (B. 1912)

"I've seen and done just about everything, it seems," says Howard Clifford. The award-winning writer and photographer, now in his ninth decade of life, is not exaggerating. Clifford has led a remarkable and richly varied career over much of the last century. But one event stands out—the run for his life on collapsing Galloping Gertie in November 1940.

His story began 28 years earlier. Born in 1912 in Wausau, Wisconsin, Howard Clifford (then known as "Howie") came west as a teenager when his parents moved to Tacoma in 1926. His interests in sports and writing appeared early. Athletic and bright, Howie enjoyed playing football and writing stories for the Stadium High School newspaper.

After graduating in journalism from the College of Puget Sound (today, the University of Puget Sound), Clifford found work in 1935 at the *Tacoma Ledger,* which later became the *News Tribune.* There his assignments covered everything from the women's page to sports. His growing interest in photography led to occasional work as a back-up cameraman for the newspaper. Just a few days before the Narrows Bridge opened in July 1940, Clifford was out taking photos. His adventuresome spirit led him to walk up one of the suspension cables to a tower, although he did wear a safety harness for the trip.

On the morning of November 7, 1940, Clifford was in the *Tribune* office when word came that Galloping Gertie was in serious difficulty. He was dispatched as the back-up cameraman along with veteran reporter Bert Brintnall. Clifford shot several photos of the span's final moments. His last picture is

blurred because the shutter snapped the same instant when the bridge started to collapse. By late afternoon, the spectacular images were sent across the country.

Clifford went on to a highly successful career, mostly as a journalist and photographer, but his many talents and ambitions led him into a host of other activities as well. He has been a commercial airline pilot, race car driver, sports announcer, film producer, public relations manager and consultant, law officer, U.S. Marine (serving during World War II in the Pacific), ski instructor, editor, publisher, and travel writer. He has photographed U.S. presidents from Franklin D. Roosevelt to Bill Clinton. Clifford's publication credits include eight books, as well as hundreds of articles and photographs. Today, Clifford resides in South Seattle.[1]

Howard Clifford, ca. 1995.
Courtesy of Howard Clifford

JAMES BASHFORD (1877–1949)

Professional photographer James "Jim" Bashford gets credit for snapping the most famous photograph of Galloping Gertie's collapse. On the morning of November 7, 1940, Bashford received news that the bridge was behaving wildly and rushed to the Narrows. He shot various photos of the twisting bridge and the first splash from debris hitting the water. Then, from the south side near the east end of the bridge, he captured the center span's collapse in an image that has been seen around the world. It continues to be the most frequently reproduced photograph of the disaster.

Bashford was born on March 19, 1877, in Boscobel, Wisconsin. His family moved to Tacoma from Iowa when Jim was two years old. He spent the rest of his life in Tacoma, watching it grow from a pioneer town to a major metropolis. He recorded important milestones of the city's history and development along the way.

In the 1890s, young Bashford worked on steamers plying waters from Puget Sound to Alaska. He soon found himself involved in what became a long career with local newspapers, including the *Tacoma Ledger*, *Seattle Times*, *Tacoma News*, *Tacoma Times*, and *Tacoma Tribune*. From 1939 to mid-1940, Bashford worked for the Thompson Photo Service, which the Washington Toll Bridge Authority had hired to document the building of the first Narrows Bridge. Bashford snapped more than 400 photographs of the construction. He also continued doing free-lance photography for Tacoma newspapers.

Bashford related many of his Galloping Gertie memories to his daughter, Ann. His story of workmen sucking lemons to combat nausea caused by the span's bouncing has become a part of the bridge's lore. Bashford died in Tacoma on July 3, 1949, at the age of 72. In 1998, Bashford's daughter and grandson donated his 4 x 5 Graflex RB-D camera, a complete set of the Bashford-Thompson photographs, and the original negative of his famous Galloping Gertie picture to the Gig Harbor Peninsula Historical Society.[2]

LEONARD COATSWORTH (1895–1956) AND TUBBY

On July 5, 1940, just four days after the Narrows Bridge opened to traffic, Leonard Coatsworth celebrated his 45th birthday. Coatsworth was a reserved, quiet man, even somewhat shy. He wore wire-rimmed glasses and usually donned a hat when outdoors. A news editor for the *Tacoma News Tribune*, the Missouri-born Coatsworth had a reputation as a nice, highly talented newspaperman. At the time, he was the only *Tribune* staff member who could cover a story, write it up and edit it, type-set the edition, and run the presses.

Leonard Coatsworth and his ill-fated Studebaker at the Arletta beach house.
Courtesy of Gerry Coatsworth Holcomb

He had many friends at the *Tribune*, where he was considered a true scholar of English and noted for his outstanding prose. Intelligent and witty, Coatsworth often was "the court of last resort," his colleagues said later, "in the frequent arguments over grammatical construction." They also referred to Leonard as "a walking dictionary. If one of the staff needed to check spelling or meaning for a word, one of them would say, 'Don't bother with the dictionary, just ask Leonard.'" He loved words and had a droll, sometimes corny sense of humor—to him puns were the purest form of humor.

Coatsworth had married Ethel Tanner in 1920. The couple had one daughter, Gerry, born in 1925. Leonard, Ethel, and Gerry often spent time at a summer home in Arletta on the Peninsula, a few miles west of the Narrows. He also occasionally invited the editorial and news staff to their summer place. Only the weekend before the collapse, the Coatsworths hosted the last picnic of the season for the *Tribune* staff.

Gerry loved dogs, especially her black cocker spaniel, Tubby. As the only victim of Galloping Gertie's collapse, Tubby has earned

Veteran newspaperman Leonard Coatsworth shortly before his retirement from the *Tacoma News Tribune* in 1956.
Courtesy of Gerry Coatsworth Holcomb

a special place in the hearts of many. Gerry Coatsworth (now Holcomb) describes Tubby as a mixed breed, but mostly cocker spaniel. She had him since he was a puppy. In 1940, Gerry was 15 years old and Tubby was about 8. He had short legs and walked with a bad limp, having been struck by a car several years earlier that left him unable (or unwilling) to put his full weight on the right forepaw.

Gerry recalls, "My mother didn't allow Tubby in the living room, but he tried to sneak in whenever he thought no one was looking. At the words, 'Back in the kitchen!' his limp became positively pathetic, as he hobbled dramatically out of the living room. At other times he got along pretty well. I was a tomboy and he was my constant companion."

After the bridge collapse, newspapers and many people blamed Coatsworth for not saving Tubby, while hailing the "courageous" effort made by Professor Farquharson. However, this assessment does not tell the complete story—Coatsworth had tried to coax Tubby from the back seat, but the terrified dog refused to move. In addition, Coatsworth needed both hands to hold on to the curb as he struggled to get off the twisting bridge. It would have been impossible to hold the panicked dog and at the same time clutch the curb to keep from being pitched into the Sound as the span tilted sideways. When Coatsworth tried to return to the car for Tubby, he was badly scraped and bruised, and in shock, and the bridge's motion had become more violent. Farquharson attempted to rescue Tubby during a lull in the oscillations, but he too had to give up and he staggered back to safety just moments before the center span collapsed. Coatsworth's colleague, photographer Howard Clifford, also tried to save Tubby, but was unable to get close to the car.

Gerry, now in her early eighties, still loves dogs. A dog behaviorist and trainer, she shares her home with three dogs and volunteers at the local humane society. Gerry clearly remembers the day she lost Tubby. "I was a senior at Stadium High School. Some girl

came up to me and said something like, 'Oh, you're the girl whose father was on the bridge when it fell down.' I ran to the office to call home, but there was a teacher on the phone. I was upset and crying, but she refused to relinquish the phone even though I begged her. I had to wait 20 minutes before she got off and I could call and find out if my Dad was alive. Finally, I learned he was alive, his knees and hands were scraped and bruised, and he was shaken up. And they told me Tubby went down with the car. I took the bus home. It was very traumatic. I loved my dog."

The loss of Tubby touched animal lovers around the country, and Gerry received more than a dozen offers for a free cocker. She chose one from Richmond, Virginia, a black and white female puppy that arrived some weeks later by rail. Gerry wanted to name her "Narrows Bridget," but her father refused, so they called her Cobina. Gerry fondly remembers Cobina. "I loved her, of course, but she really became my parents' dog after I left for college less than a year later, in 1941, and they grew

Tubby, as remembered by Gerry Coatsworth.
Drawing by Mason Holcomb

Gerry Coatsworth with Cobina, the puppy she received after the loss of Tubby.
Courtesy of Gerry Coatsworth Holcomb

to love her as well. Dad always insisted she barked with a Southern accent."

The exact location of Coatsworth's car in the Sound remains unknown. Coatsworth had stopped approximately 450 feet west of the east tower. The distance at mid-span from the roadway to the water was about 190 feet, although the road was tilting wildly at the time of the collapse, affecting that measurement by up to 28 feet. The water depth near the east pier where the automobile fell and sank is about 125 feet. The swift tides apparently have swept the vehicle away from Galloping Gertie's ruins.

Coatsworth had trouble getting reimbursed for the loss of his 1936 Studebaker. The day after Galloping Gertie's collapse, he sent in a claim, but six months passed with no news and no money. Finally, Coatsworth wrote a letter to the Washington Toll Bridge Authority, asking for a response. The WTBA replied that they had not received his first request. Coatsworth quickly discovered that his attorney had never followed up on the claim. With a new attorney, the process moved on, though not quickly. Finally, in December 1941—more than a year after the collapse— the WTBA reimbursed Coatsworth $450.00 for the loss of his car and, in a separate action, $364.40 for the loss of his vehicle's "contents," including Tubby. The WTBA, however, refused to refund the 50¢ toll that Coatsworth had paid before driving onto the bridge that day. "Dad thought this was pretty funny," says Gerry. "He used to joke that they owed him a free trip on the new bridge."

The *Tacoma News Tribune* erroneously hailed Coatsworth as the "Last Man on the Bridge" when they published the account of his escape. Wire services around the country picked up the story and Coatsworth became an instant celebrity. He traveled to New York for radio, newspaper, and magazine interviews, and later to California for a television appearance. Coatsworth retired from the *Tribune* in 1956. He was 61 years old and seemingly in good health. He and Ethel, with Cobina, planned to move to Santa Barbara, but a fast-acting kidney ailment intervened and he died in November 1956.[3]

HALLETT R. FRENCH (1895–?)

The news of Galloping Gertie's demise was especially shocking to Hallett R. French. The 45-year-old Seattle-based insurance agent for Merchants' Fire Assurance Company of New York had authority to write policies without prior approval and had made the most of it. French had written an $800,000 policy on the bridge for the State, believing there was no chance of the bridge collapsing. Rather than reporting the transaction and turning over the $70,000 premium to the firm's main office, French deposited the money in his bank account. He was vacationing in Idaho when he received word on November 8 that the bridge was gone.

On December 2, 1940, Seattle police arrested French for grand larceny. He was denied bail and a trial was set for February 1, 1941. French returned some $17,500 of the missing funds to his company. His businessmen friends in Seattle begged the court for leniency. On February 7, French pleaded guilty and the judge promptly sentenced him to 15 years in the Washington State Penitentiary in Walla Walla.[4] He served only two years and was released for good behavior. He soon found a job at a Seattle shipyard, and passed into obscurity as newspapers stopped printing stories about him.

FREDERICK B. "BURT" FARQUHARSON (1895–1970)

As a professor of civil engineering at the University of Washington from 1925 to 1963, Farquharson left a lasting legacy in bridge engineering with his aerodynamic studies of the 1940 and 1950 bridges.

Born in Boston, Massachusetts, in 1895, Farquharson served in World War I with the Canadian Army and the Royal Air Force. The

Germans captured him in 1917 and Farquharson spent the last 15 months of the war in a prisoner of war camp. Upon returning to the United States, Farquharson attended the University of Washington, graduating in 1923. After two years working for the Boeing Company, the able young engineer accepted an offer to join the University of Washington faculty. He went on to head the Engineering Experiment Station and became a world-recognized authority on aerodynamic testing for bridge designs. The pipe smoking, mustachioed professor was widely appreciated for his quiet, kindly, and efficient demeanor.

Farquharson stood on the 1940 bridge the day that it collapsed, intently monitoring its behavior, snapping photos, and taking motion picture film. The movie credited to Farquharson remains a classic that is viewed by engineering students around the world.

In the 15 years that followed, Farquharson's pioneering aerodynamic studies were applied to the building of the 1950 Narrows Bridge and other suspension spans around the world. He retired from the University of Washington in 1963, and died at the age of 75 on June 17, 1970.[5]

Charles E. Andrew (1884–1969)

Charles Andrew was one of Washington's most visionary and controversial engineers. A soft-spoken man with great energy, confidence, and determination, he served as chief consulting engineer for the Washington Toll Bridge Authority for two decades. During his tenure, Andrew guided construction of the 1940 and 1950 Narrows bridges, the first two Lake Washington floating bridges, and the Hood Canal Floating Bridge. To his admirers, Andrew was "a genius," but others criticized him for building "blow away bridges."

Charles Andrew was born in Illinois in 1884 and graduated with a bachelor's degree in civil engineering from the University of Illinois in 1906. He immediately moved to Oregon, where he began his career building

railroad bridges. In 1921, Andrew moved to Washington and served for the next six years as the first bridge engineer for the State Highway Department. He left in 1927 for the San Francisco area, where he remained for the next 10 years. In 1931, Andrew became the principal engineer on the San Francisco-Oakland Bay Bridge, leading the design and construction of the world's largest bridge, completed in 1936.

When he returned to Washington in March 1938, Andrew was a celebrity in engineering circles. He immediately set to work on the Tacoma Narrows Bridge project as chief consulting engineer for the WTBA. When the bridge collapsed in November 1940, some observers blamed Andrew. In fact, Andrew had fought hard against the cost-cutting design compromises that led to the structure's collapse. The second Narrows span, designed to Andrew's standards and specifications, has now withstood Northwest storms for half a century.

Arnie Colby, one of the men who worked on the 1950 bridge and knew Andrew, recalls him as "a very nice man. He was wealthy from playing the stock market. He was on the telephone a lot, talking to his broker in New York. He was quiet, but very determined, and extremely bright. He knew every detail about the bridge. You never had to worry about whether he understood you."

Andrew also led the design effort for the 1961 Hood Canal Floating Bridge and the 1963 Evergreen Point Floating Bridge, the largest concrete floating bridge in the world. When the Hood Canal Bridge partly sank in a 1979 storm, critics targeted Andrew, despite the fact that he had died 10 years earlier. As with the 1940 Narrows Bridge, the criticism was unjust. Andrew had opposed the design compromises forced by budget limits, and an investigation revealed that the bridge might not have failed if pontoon hatches accidentally left open had been properly closed.

Charles Andrew's reputation in recent years has begun to regain its former stature. His

Charles Andrew, the principal consulting engineer for the Washington Toll Bridge Authority, ca. 1940.
WSDOT

unrealized visions included a tunnel crossing under Puget Sound from Seattle's Alki Point to Bainbridge Island, a floating bridge to Vashon Island from Fauntleroy, and a large suspension bridge over Colvos Passage.[6]

DEXTER R. "DEX" SMITH (1891–1973)

In 1941, WTBA's principal consulting engineer, Charles Andrew, handpicked Dexter "Dex" Smith to design the 1950 Narrows Bridge, a daunting assignment following Galloping Gertie's well publicized collapse. Smith had been a chief bridge design engineer for the Oregon State Highway Department.

Smith was born in 1891 near Portland, Oregon. In 1914, he graduated from Oregon State College (now Oregon State University) and began his career there as an instructor, teaching civil engineering until the end of the 1920s. In 1929, Smith moved to the Oregon State Highway Commission, where he designed bridges for the next nine years. At that time, he became a close associate of the well-known Oregon State Engineer, Conde B. McCullough, a recognized expert in designing short-span suspension bridges. Smith's accomplishments in structural bridge design in Oregon soon earned him a growing reputation.

It was while working as a private consulting engineer that Smith was contacted by Andrew, who requested his help in designing the replacement bridge at the Narrows. The talents that Smith brought to the job soon became evident. He led the State's design team through exhaustive testing, collaborations with University of Washington researchers, financial delays, and political hassles from 1941 until the successful completion of the bridge in 1950. Smith was so meticulous that he personally made all of the final design drawings himself. Today, the Tacoma span's strength, stability, and durability are testimony to Smith's skill and determination.

After 1950, Smith served as a design engineer consultant for a suspension bridge over Chesapeake Bay in Maryland that was completed in 1952. Next, he decided to return to his old job at the Oregon State Highway Department and also taught structural engineering at Oregon State University. A colleague remembers Smith as a short man with a "feisty" temperament. However, Smith was endlessly patient with his students and provided a careful guiding hand that helped them to find answers for themselves. Apparently, he took little interest in his appearance. Jackson

Dexter R. Smith, design engineer for the 1950 Narrows Bridge, ca. 1957.
Oregon Department of Transportation

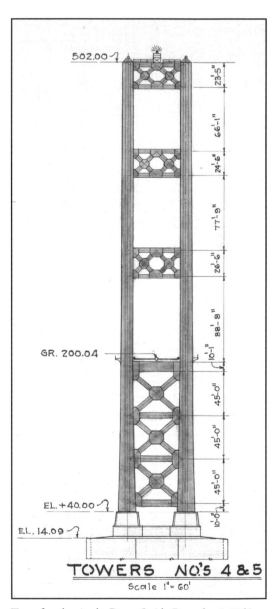

Tower face drawing by Dexter Smith, December 5, 1945.
WSDOT

Durkee, a Field Engineer for Bethlehem Steel, remembers the day when Smith came into his office dressed in a shabby overcoat. "He looked so dowdy," Durkee recalled, "at first I thought he was somebody there to empty the garbage!" Dexter Smith retired from the Oregon State Highway Department in 1957, and passed away in November 1973 at age 82.[7]

Charles "Chuck" Munson, a former student of Smith's at OSU, worked under him at the Oregon Highway Department. In recalling Smith's career, Munson remembers a fascinating incident. In 1939 or 1940, only months before Galloping Gertie failed, Smith reviewed Leon Moisseiff's design for the Tacoma Narrows Bridge. Smith presented a paper on his findings at a national engineering conference, probably held by the American Society of Civil Engineers. Attending the session were some leading suspension designers of the day, including Joseph Strauss, David Steinman, and Leon Moisseiff. Smith predicted that under specific conditions—a wind of 40 mph striking the bridge's solid plate girder at an angle of 35 degrees—the span would collapse. When Smith finished, the stunned audience sat silent. Then they rose to their feet, shouting and demanding that Smith "get out."

Another of Smith's contemporaries, James Howland of Corvallis, Oregon, repeated this story. Howland added other details, but the essence of the episode is the same. However, because records of the 1939 and 1940 American Society of Civil Engineers national meetings are unavailable (or were not kept),

printed documentation of this dramatic incident apparently does not exist. Smith's critique of Moisseiff's design, however, certainly provides a logical reason for why Charles Andrew selected Smith to design Galloping Gertie's replacement.[8]

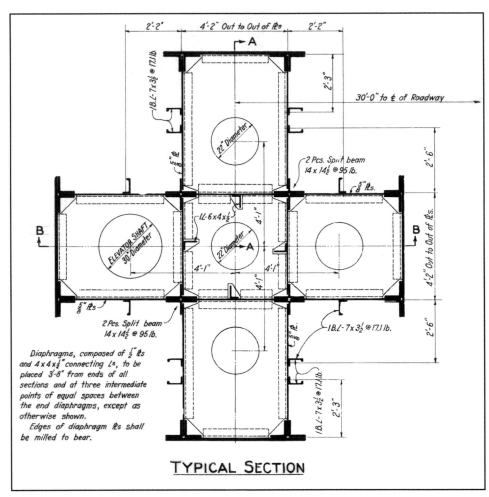

Tower leg cross section, December 5, 1945. *WSDOT*

Notes

1. Howard Clifford, interviews, October 2003 and September 2005; Howard Clifford, "Cameraman Was on Cracking Span," *Tacoma News Tribune*, November 8, 1940; Howard Clifford, "A Day to Remember; When the Tacoma Narrows Bridge Fell," TAPCO newsletter, undated copy in Tacoma Narrows Bridges History files, Gig Harbor Peninsula Historical Society; "Lensman Recalls Gertie's Gallop," *Tacoma News Tribune*, November 7, 1975; "Last Man on Bridge Recounts Gertie's Collapse," *Tacoma Sun*, November 2, 1980; Howard Clifford, "Memories of Gertie," *Tacoma News Tribune*, November 1, 1990.

2. Bashford information provided by Vicki Blackwell, Gig Harbor Peninsula Historical Society.

3. Gerry Coatsworth Holcomb, telephone interview, January 7, 2006; Gerry Coatsworth Holcomb, e-mails to the author, January–September 2006; "Leonard B." (editorial), *Tacoma News Tribune*, November 27, 1956; "Phobia Saved Newsman's Life," *Tacoma News Tribune*, November 7, 1975; Howard Clifford, interview, October 2003; letters to and from Leonard Coatsworth to the Washington Toll Bridge Authority and others, May 6, 1941, to December 18, 1941, in "Coatsworth" folder, Box 42, WTBA, WSA.

4. "15 Years for Span Insurer," *Tacoma News Tribune*, February 10, 1941.

5. "F. Burt Farquharson, Bridge Engineer, Dies," *Seattle Times*, June 18, 1970.

6. "Charles E. Andrew, A Builder of Bridges and Men; Documentation Submitted to the American Society of Civil Engineers for Consideration of the Ernest E. Howard Award Committee for the Year 1960," Seattle Section, ASCE, January 1960, copy in Biography File, Manuscripts, Special Collections and University Archives, University of Washington Libraries; Ross Cunningham, "Unfair to Make Him Scapegoat," *Seattle Times*, June 8, 1979; "Designer of Floating Bridge Succumbs," *Daily Olympian*, July 6, 1969; "Rites Planned for Charles Andrew, 85," *Seattle Times*, July 5, 1969.

7. "Designer of Bridges Dies," *Oregon Statesman*, November 20, 1973; (editorial) *Oregon Journal*, September 5, 1935; (obituary) November 22, 1973; (retirement) October 25, 1957; *Oregonian*, October 25, 1957, and November 21, 1973; e-mail to the author from Larry Landis, Archivist, Oregon State University, November 18, 2003; Robert W. Hadlow, *Elegant Arches, Soaring Spans: C.B. McCullough, Master Bridge Builder* (Corvallis: Oregon State University Press, 2001), 39, 90, 92, 114, 121, 150; Lewis Melson, telephone interview, February 16, 2004; Jackson Durkee, telephone interview, November 8, 2005. See the Bibliography for Dexter Smith's publications published jointly with Conde B. McCullough and Glenn S. Paxson.

8. Charles "Chuck" Munson, telephone interviews, November 2003 and January 2004; James Howland, telephone interview, February 16, 2004. Charles Munson and James Howland are unacquainted and were separately informed about Smith's critique of Moisseiff's design.

Chapter Twelve

Bridge Connections, 1950 to the Present Time

Opening Day for "Sturdy Gertie"

On October 14, 1950, an inaugural celebration capped the completion of Galloping Gertie's replacement. The new span, nicknamed "Sturdy Gertie" by local promoters, was wider (for four lanes of traffic), heavier, and stronger than its predecessor. The span was a landmark of aerodynamic bridge engineering and represented a new era in suspension design.

On that brisk October day, a large crowd attended the gala event, including Governor Arthur B. Langlie and other dignitaries at the podium. One speaker told the large crowd, "You see before you a memorial to indomitable courage and will power, as well as a monument to vision, research, design and skill." A souvenir booklet published for the grand opening proclaimed, "A combination of men's dreams, fortitude, and inventive ingenuity, which with private capital has created a

Opening day at the Tacoma Narrows Bridge, October 14, 1950.
WSDOT

Scouts parade across the new bridge during opening day ceremonies.
Tacoma Public Library 6656

**STATE OF WASHINGTON
DEPARTMENT OF HIGHWAYS**

TACOMA NARROWS BRIDGE TOLL SCHEDULE
Effective March 1, 1953

Class	Type—		Toll
1.	Automobile and driver		$0.50
2.	Automobile and driver, plus 1 passenger		.60
3.	Automobile and driver, plus 2 passengers		.70
4.	Automobile and driver, plus 3 passengers		.80
5.	Automobile and driver, plus 4 passengers		.90
6.	Automobile and driver, plus 5 passengers		1.00
	(Ambulance, hearse, station wagon, trucks, and trucks licensed under 4000 lbs. included in above types) (Note 1)		
7.	Truck, gross license weight 4,000 lbs. to 6,000 lbs.		.65
8.	Truck, gross license weight 6,001 lbs. to 12,000 lbs.		1.00
9.	Truck, gross license weight 12,001 lbs. to 18,000 lbs.		1.50
10.	Truck, gross license weight 18,001 lbs. to 24,000 lbs.		2.00
11.	Truck, gross license weight 24,001 lbs. to 30,000 lbs.		2.50
12.	Truck, gross license weight 30,001 lbs. to 36,000 lbs.		3.00
13.	Truck, gross license weight 36,001 lbs. to 42,000 lbs.		3.60
14.	Truck, gross license weight 42,001 lbs. to 48,000 lbs.		4.20
15.	Truck, gross license weight 48,001 lbs. to 54,000 lbs.		4.90
16.	Truck, gross license weight 54,001 lbs. to 60,000 lbs.		5.60
17.	Truck, gross license weight 60,001 lbs. to 66,000 lbs.		6.30
18.	Truck, gross license weight 66,001 lbs. to 72,000 lbs.		7.00
	(Note 2)		
19.	Bus, through, chartered or suburban, including driver		1.00
	Passengers in bus, each (Note 8)		.10
20.	Special, unclassified units, toll to be determined at time of passage.		
20.	Vehicle, horse drawn, including driver		.75
20.	Horse and rider		.25
21.	Extra passengers in vehicle other than bus, each (Note 8)		.10
22.	Auto trailer, one or two wheels, not house trailer or for livestock		.25
22.	Auto trailer, other than provided for above		.55
23.	Motorcycle, with or without sidecar, including driver		.25
24.	School bus, including driver		.75
	Passengers in school bus, each		.05
25.	Pedestrians or bicycle and rider (Note 8)		.15
26.	Commute book, auto and driver, 40 crossings in 30 days (Notes 3 and 7)		12.00
27.	Commute book, passenger or pedestrian, 40 crossings in 30 days (Note 3)		3.00
28.	Commute book, public agency auto and driver, 25 crossings (Note 5)		7.50
29.	Commute book, public agency truck and driver, 25 crossings (Note 5)		31.25
30.	Commute book, public agency passenger or pedestrian, 30 crossings (Note 5)		2.25
31.	Commute book, school child, 50 crossings (Note 4)		2.50
	(Over)		

Toll schedule, March 1953.
Washington State Archives

masterpiece of engineering skill together with the solution of a dire economic need."[1]

Finishing details took another year. Not until November 1951 was the bridge officially completed. As of June 30, 1952, Charles Andrew reported that construction costs totaled $14,011,384. Like its predecessor, the 1950 Narrows Bridge became the world's third longest suspension bridge. Today it ranks as the fifth longest in the United States.

POPULAR AND SUCCESSFUL

The new bridge delighted area residents. After the opening day ceremonies, the first person over the span was a truck driver, who was "thrilled" to pay the toll. The fee for a car and driver was 50¢ each way, or $1 roundtrip. Passengers were an additional 10¢. The same trip by ferry had cost $7.

The revenue collected on the first day of operations totaled a colossal $11,541. Daily records remained astonishingly high for more than 14 years, until the tolls were removed. As with Galloping Gertie, the new bridge proved far more popular than traffic surveys had predicted. The average daily traffic count rose steadily. In the first months, October to December 1950, the toll span averaged 3,904 vehicles a day. By 1955, the number was up to 4,699.

In 1960, the daily traffic averaged 6,218 vehicles, and over the next five years almost doubled, reaching 11,267 in 1965. Total revenues then stood at $19 million, substantially beyond the construction bond issue's $14 million and accrued interest. Citizens anxious to see the tolls removed finally convinced the government that the time had come in 1965. A bill to remove fees from the Narrows Bridge passed the State Legislature.

On May 14, 1965, jubilant bridge users celebrated. In a festive ceremony at the toll plaza, Governor Dan Evans signed legislation removing the WTBA tolls 13 years ahead of schedule—resounding proof of the bridge's value and popularity. Since then, the Narrows Bridge has been owned and maintained by the Washington Department of Highways, and its successor, the Washington State

Department of Transportation (WSDOT). Workmen soon removed the booths and gates

Governor Dan Evans hosts the celebration for the removal of tolls, May 1965.
Washington State Archives

WAR MEMORIAL, 1952

Civic leaders led a community effort to establish a "Living War Memorial Park" at the east end of the bridge. The steering committee included representatives from almost every civic, service, fraternal, and military organization in Tacoma. The dedication ceremony on Armistice Day, November 11, 1952, honored the sacrifices of Tacoma area men and women who served in the armed forces. Volunteers established the 1½-acre

Sailors at the Living War Memorial Park, October 1952.
Tacoma Public Library 22808

park that included a lawn and an 800-pound bronze bell. Farmers donated flower bulbs, U.S. sailors assigned to the Puget Sound region helped build a parking lot, and the Tacoma Bricklayers and Hod Carriers' Union erected the 20-foot stone monument. The site was deeded to the city by the Young Men's Business Club, which had been given the property by Harold A. Woodworth. (The memorial was removed for the construction of the 2007 bridge, but with plans to reinstall it after completion of the new span.)

Motorists stop for a traffic survey, ca. late 1950s.
WSDOT

the Peninsula. Traffic growth over more than a half-century has clearly reflected an increase in population, tourism, government facilities, and commercial ventures. By 1970, the daily average number of vehicles climbed to 21,164. A decade later, it had nearly doubled, and by 1990, the daily count was 66,573. In 2000, more than 32 million cars and trucks crossed the bridge, averaging 88,000 a day.

Currently, the daily count can be as high as 120,000 vehicles on some of the busiest days. Congestion can be a problem, with rush hour gridlock, accidents, and stormy weather causing bumper-to-bumper traffic jams that back up for miles.

OCTOPUS WRESTLING: END OF ANOTHER ERA

One the world's largest octopus species thrives in Puget Sound and the Narrows. Titlow Park, just south of the 1950 Narrows Bridge, is a popular starting point for scuba divers visiting Galloping Gertie's sunken ruins. From the late 1950s to 1965, Gertie's wreckage provided the backdrop for fans of what Tacoma promoters called the "World championship of octopus wrestling."

It was an odd "sport." Each April, divers gathered at Titlow Beach near the old Sixth Avenue dock. A gun shot launched dozens of suited, masked, and fin-clad swimmers into the waters. Some used only snorkeling gear,

LOVE AND THE BRIDGE

People love the Tacoma Narrows Bridge for many reasons. One couple decided it was the perfect place for a wedding. On a brisk Friday morning at the end of October 1998, commuters crossing the bridge might have noticed the bride and groom. Airman First Class Bobby Collins wore his uniform, and optician April Koons looked lovely in her dark blue, ankle-length dress. Both 22 years of age, they became husband and wife at a brief ceremony. Why the Narrows Bridge? Why not? "That sounds cool," said Collins when the idea came up. After the ceremony, he told a local newspaper reporter, "It's the greatest feeling in the world."[3]

(although the toll houses still stood at the east end of the span).[2]

However, not all of the Peninsula's residents and commuters were excited to see the tolls end in 1965. Some citizens feared this would prompt a virtual "invasion" by newcomers into their small communities. They fought to keep the tolls, but were unsuccessful.

IMPACT ON TACOMA, THE PENINSULA, AND GIG HARBOR

The 1950 bridge, of course, became a vital link between Tacoma and communities on

TRAFFIC 1950–2000

Year	Annual number of vehicles	Daily average
1950 (ferry)	593,871 (Jan. 1–Oct. 13)	2,076
1950 (bridge)	280,464 (Oct. 14–Dec. 31)	3,550
1960	2,269,570	6,218
1965	4,112,455	11,267
1970	7,724,860	21,164
1980	14,225,145	38,973
1990	24,299,145	66,573
2000	32,120,000	88,000

Toll plaza at the east entrance, 1951.
Washington State Archives

Year	Tacoma[a]	Gig Harbor[b]	Bremerton[c]
1940	182,081	770	44,387
1950	275,876	803	75,724
1960	321,590	1,094	84,176
1970	411,027	1,666	101,732
1980	485,643	2,429	147,152
1990	586,203	3,236	189,731
2000	700,820	6,465	231,969

a. Tacoma's Principal Metropolitan Statistical Area includes parts of Pierce County.
b. Gig Harbor was incorporated in 1946.
c. Bremerton's Principal Metropolitan Statistical Area includes parts of Kitsap County.

while others donned scuba equipment to stay underwater longer. Heats of three-man teams competed for the "world octopus wrestling" crown by landing the most pounds of octopuses. Periodically, they rose to the surface holding a mass of wiggling tentacles and dumped the catch on a scale for weighing. The winners received a cash prize for the highest combined octopus weight, and most of the hapless animals were returned to the Narrows.

In the late 1950s, the contests attracted a couple hundred onlookers. The largest catch in 1957 weighed 80 pounds. By the early 1960s, as many as 3,000 spectators watched more than a hundred divers from Washington, British Columbia, Oregon, and California go after the eight-armed creatures. In 1964, contestants gathered 24 octopuses, weighing a total of 700 pounds. That year the near-winner was a diver who tussled with a 102-pound octopus, but the animal died before it could be weighed. The winner tipped the scale at a much lighter 58 pounds. The largest of the group ended up at the Seattle Marine Aquarium, and the remainder were dumped back into the Sound.

By then, the once-feared "monsters of the deep" were appreciated as timid, curious, friendly, and intelligent creatures, but their numbers and average size in Puget Sound had dwindled alarmingly, mainly due to pollution. In 1965, state law forbad harassment of undersea animals and imposed a $1,000 fine and a one-year jail sentence for anyone convicted of

octopus wrestling. Most serious divers had left the animals alone for years and were pleased to see the contest end. As the "world championship" event neared that year, its sponsor, the Puget Sound Mudsharks Diving Club, withdrew support, ending the so-called "sport" at the Narrows.[4]

GERTIE TODAY: HISTORIC RUINS AND ENVIRONMENTAL ASSET

Galloping Gertie's steel and concrete retain a special place in history. The great span's ruins extend more than 2,000 feet between the east and west piers, covering some 20 acres of sea floor. In 1992, the wreckage was placed on the National Register of Historic Places, as well as the Washington State Register. The effort was unprecedented—it was the first National Register nomination relying on sonar imagery for documentation.

The nomination summarized Gertie's significance: "The collapse of the Tacoma Narrows Bridge was a hallmark in the history of bridge design and civil engineering. Its impact is still felt in the profession today. The bridge's remains at the bottom of Puget Sound are a permanent record of man's capacity to build structures without fully understanding the implications of the design and the forces of nature."[5]

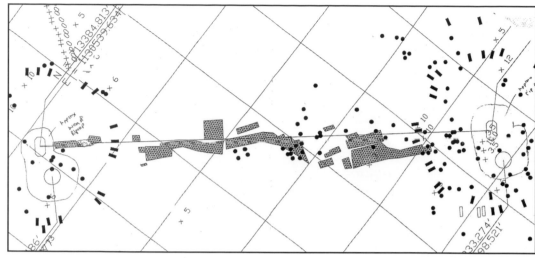

One of the driving forces behind the effort was Robert "Bob" Mester, president of Underwater Atmospheric Systems Inc. (UAS). Using sonar soundings, Mester and UAS experts charted the resting places of the center span and other debris from Gertie's collapse. The site plan and other information from UAS added vital knowledge about the ruins and augmented its official recognition as one of the country's historic treasures. The effort also contributed to a 30-minute documentary "Gertie Gallops Again," co-produced by Bob Mester and Tacoma Municipal Television in 1998. The film aired on TMT's *CityScape*, winning the 13th Annual Award for Cable Excellence for the best documentary done in Washington.

Today, Galloping Gertie is one of the world's largest man-made reefs. Its environmental significance is constantly taking on new meaning. The abundance of marine life that thrives among the ruins is one of Washington's great natural assets, but it is a fragile ecosystem. Before Gertie, the floor of the Nar-rows was a harsh habitat for most marine life, with treacherous currents ripping through the channel four times a day, scouring the bottom. The dropping of 600-ton anchor blocks in 1939 began to change that. Then, when Gertie collapsed, the massive girder sections, concrete rubble, and a half-mile of roadway significantly changed the natural environment. The debris created an underwater haven from the swift tides.

A rich diversity of marine life now thrives among and near the ruins, including starfish, anemones, mussels, piddock clams, barnacles, crabs, octopus, flatfish, sculpins, rockfish, lingcod, wolf eel, and surf perch. The Narrows also is a natural migration corridor and feeding habitat for a great variety of fish, marine mammals, and waterfowl, including salmon, dogfish shark, skates, black rockfish, the tiny Pacific sand lance, sea lions, and diving birds that plunge to a depth of 140 feet. The old span, looming ghost-like in the dark undercurrents of the Narrows, remains as both a historical artifact and an environmental legacy.

Notes

1. "Narrows Span Opened as Cannons Boom," *Seattle Times*, October 14, 1950; "Souvenir of Tacoma Narrows Bridge, Tacoma, Washington," dedication program, Pioneer Incorporated, Tacoma, copy in WSDOT records, Washington State Archives.

2. "Toll Bridges and Ferries: Tacoma Narrows Bridge," Washington State Highway Commission First Biennial Report; Washington Department of Highways Twenty-fourth Biennial Report, 1950, 1952 (Olympia: State Printing Plant [1952]), 63–64; "Ceremony Tomorrow to Celebrate Narrows Bridge as Toll-Free Span," *Seattle Times*, May 13, 1965; Ross Cunningham, "Western Sound Area Just as Entitled to Free Bridges as Rest of State," *Seattle Times*, December 9, 1962; "End of Tolls Sought for Narrows Span," *Seattle Times*, March 26, 1964; "House Unit Would Halt Narrows Tolls," *Seattle Times*, February 13, 1965; "Narrows Bridge Toll Removal Sought in

Court," *Daily Olympian*, March 26, 1964; "Narrows Bridge Tolls to Stay," *Daily Olympian*, February 6, 1963; "Narrows Bridge Top Money-maker," *Daily Olympian*, January 15, 1964.

3. Khris Sherman, "Sky-High Couple Wed on the Bridge," *Tacoma News Tribune*, October 31, 1998.

4. Joe Contris, "Octopus Wrestling, Anybody? Lots of Fun," *Tacoma Sunday Ledger-News Tribune*. August 11, 1957; "Seattle Octopus Wrestling Team Keeps World Title," *Tacoma News*

Tribune, April 22, 1963; Don Hannula, "58 Pound Octopus Brings Diver First Place in Colorful Grappling Contest at Titlow Beach," *Tacoma News Tribune*. April 19, 1964; Roland Lund, "Forfeit? King Octopus Keeps Crown," *Tacoma News Tribune*, March 28, 1976.

5. National Register of Historic Places nomination form for the Tacoma Narrows Bridge Ruins/Galloping Gertie, Pierce County, Washington, 1992.

View of the east anchorage and toll plaza from Tower #5.
WSDOT

CHAPTER THIRTEEN

THE BRIDGE MACHINE SINCE 1950

HOW BIG IS THE BRIDGE?

The two towers of the 1950 Tacoma Narrows Bridge rise 467 feet above the piers and 507 feet above the mean high-water level (as measured to the centerline of the cables at the tops of the towers). By comparison, the Seattle Space Needle is 138-feet taller at 605 feet, while the Statue of Liberty is only 305-feet high, or 162-feet shorter than the bridge's towers.

The bridge measures 60 feet between the main suspension cables and is 5,979 feet long—it would take nearly 10 Seattle Space Needles end-to-end to match that length, or nearly 20 Statues of Liberty.

MAINTAINING A MILE-LONG MACHINE

From its opening in October 1950 until May 1965, the Narrows Bridge was a toll facility. The Washington Toll Bridge Authority operated and maintained the structure, and collected more than $19 million from motorists. Since then, the bridge has been the responsibility of the Washington Department of Highways, and its successor, the Washington State Department of Transportation (WSDOT).

Looking southwest in the late afternoon.
Washington State Archives

There is more to the "housekeeping" of this complicated assemblage of machinery than one might think. Not long after the bridge's opening, inspectors discovered that the lower bolts in the cable bands were rusting. The caulking done during construction had sealed water in, rather than keeping it out. In 1954,

1950 TACOMA NARROWS BRIDGE COMPARED WITH OTHER STRUCTURES

	Total length or height	Number that would match the Narrows Bridge's 5,979-foot length
Seattle Space Needle	605 feet	9.9
Golden Gate Bridge	6,450 feet	0.9
Empire State Building	1,453 feet	4.1
Statue of Liberty	305 feet	19.6
Eiffel Tower	1,063 feet	5.6
Titanic	885 feet	6.8

Toll Plaza, Tacoma Narrows Bridge. 9-14-'51.

View of the landscaping and toll plaza at the east anchorage.
WSDOT

a major overhaul was completed under the guiding eye of maintenance engineer Harry Cornelius. All 2,200 of the bolts in the cable bands that clamped the castings securely to the suspender cables had to be removed, cleaned, painted, and reinstalled—one at a time.

Veteran bridgemen Earl White, Tommy Myers, Joe Dowsette, and Art Knoll worked on the project for nearly a year. They alternated roles, with one man staying in the shop while the other three did the removal and

reinstallation. The job was not for the butterfingered or fainthearted. To remove each of the $13 bolts, three men worked on a tiny platform hanging from one of the main suspension cables, hundreds of feet above the Narrows. Two of them handled a specially-made, four-foot-long wrench, with one man holding the working end of it in place over the bolt head, while the other man exerted all his weight on the opposite end. For reinstallation, each bolt had to be tightened, again with a man jumping on the wrench end, until it stretched 27-thousandths of an inch, measured by a micrometer.[1]

In modern times, workers are always safely attached to the hand rail cable with a large metal clip on a lanyard, connected to their harness. In the days before strongly enforced safety regulations, however, bridgemen often ignored the use of safety harnesses, or they walked along cables unclipped.

That caution-to-the-wind spirit lived on into more recent years, despite increasing federal and state safety regulations. One former maintenance employee recalls how he and a co-worker would literally race down the main cables. They wore a safety harness, but simply slid the clips along the top of the hand rail cable, instead of properly attaching them to

Bridge Maintenance and Related Expenditures, 1950–1964

	Dates	Total expenditure
General maintenance	1950–51	$4,395
General maintenance	1951–53	$37,145
Cleaning and painting towers, electrical work, west approach gully drain	1953–55	$140,470
Toll plaza vents, cleaning and painting, electrical, wiring, fender investigation, and engineering	1955–57	$47,547
Toll plaza vents, cleaning and painting, repair jacks, structure and fender repair	1957–59	$691,412
General maintenance	1959–61	$56,062
General maintenance	1961–63	$88,796
Cleaning and painting	1963–Nov. 1964	$144,972

the cable. It never started out as a race, but with one man on the north cable, one on the south, and both headed toward an anchorage, they would eye each other. Suddenly, one would pick up a little speed, and then in the next instant, they both would think, "He's not going to beat me." A second later, clips off, down they ran. Earl White proudly states, "I was the second fastest guy across the Narrows on that cable. Only one guy, Joey Orlando, could beat me."

Today, the man in charge of keeping the bridge in top shape is Kip Wylie. As maintenance supervisor, Wylie oversees three painters and two ironworkers. The bridge now is well over a half-century old, so the crew stays busy with replacement and repair work. Much of the maintenance is weather-dependent, especially the painting. The painting season extends roughly April through October. During the winter months, the crew makes sure that the bridge is free of ice and snow so that commuters have a safe and speedy transit over the Narrows.

Today, the maintenance budget is around $800,000 a year. Although that seems like a substantial sum, it goes quickly. The Narrows Bridge is a mile-long machine with many moving parts, and it takes a major effort to keep it clean and in good working order. Good management by Wylie and his crew gets the most done for the least cost.

Painting and roadway repairs absorb most of the maintenance funds. One of the bigger costs comes in repairing the 50-year-old expansion joints. Considerable movement occurs at the various expansion joints on the towers and anchorages, and especially where the east deck and west deck meet. Each end of the bridge can move up to 16 inches, expanding during hot summer weather and contracting on cold winter days. Normally, motorists never notice the expansion or contraction, but the expansion joints take a lot of punishment from temperature variations and traffic. When the Narrows Bridge speed limit was raised from 45 to 55 mph in the mid 1990s,

East anchorage, 1952.
WSDOT

the wear and tear on the expansion joints increased significantly, as did repair costs.

Painting is the most dangerous work. The bridge crew closely follows safety requirements and procedures, with each worker wearing a full-body harness in the event of an accidental fall. Workers are "hooked off" at all times with a 5,000-lb. test rope, and the harness is designed so that a person can hang suspended for several hours while waiting for rescue. In the 1950s, painters and maintenance men wore a simple belt that gave them a hanging time of only about 15 minutes before their circulation would be cut off.

Some call the bridge's color "Gertie Green," but officially it is "Narrows Green," a grayish-green shade. Today, most surfaces on the span have thick coatings of paint—at least four layers have been added since the bridge opened in 1950. The same color was used on the 1940 Narrows Bridge, although a newspaper report back then called it "Chrome Green." Ten years of working in springtime and summer (roughly April 1 to November 1) and 6,000 gallons of paint are required for the crew to completely coat the bridge.

INSPECTION AND PRESERVATION

Bridge inspections are a vital part of the maintenance work and crucial to the span's service and longevity. The maintenance crew always is looking for signs of trouble. Every Friday, two crewmembers make a general inspection tour of the towers, cables, and anchorages. Once a month, they carefully inspect the main cables for paint cracks or other problems.

Inspections by private consultants are equally important. In 1983, an independent engineering consulting firm completed a comprehensive evaluation of the bridge, making several discoveries—the suspender cables had stretched over the years, the legs of the towers showed signs of increasing stress, and the hydraulic dampers were not working well because the seals were deteriorating, causing a loss of hydraulic fluid and pressure. Another independent inspection study in 1991 used computer models and other sophisticated methods to evaluate the bridge's condition. The tests showed that the stretching of the suspender cables discovered in 1983 was not significant. But importantly, the 1991 inspection recommended a series of seismic tests to determine how well the structure would survive a major earthquake.[2]

Another study in 2000 marked a major milestone for the bridge and its maintenance crew. With plans moving forward for a second Narrows bridge (to be completed in 2007), WSDOT needed to determine the 1950 span's overall condition. A private company that specialized in reviewing the serviceability of bridges across the country performed the inspection. At the end of the intensive evaluation, the verdict was impressive—"This bridge is one of the best in the nation for its maintenance and condition."

WSDOT's Bridge Preservation Office plays a key part in maintaining the long-term serviceability of the bridge. The office performs careful inspections of all public highway bridges in Washington in accordance with national mandates and standards. The effort dates from the collapse of Ohio's Silver Bridge in 1967, which killed 46 people and drew widespread attention to the status of the nation's bridges. In response, Congress empowered the Federal Highway Administration to issue National Bridge Inspection Standards and establish a program for regular and comprehensive evaluations of all bridges that were part of the federal highway system. In 1976, these requirements were extended to all public bridges.

Since then, WSDOT has participated in the national bridge inspection program. At least once every two years, all of Washington's publicly-owned bridges are reviewed and inspected. Annually, the Bridge Preservation Office sends a report to the FHWA for its national database, the National Bridge Inventory. Sharan Linzy, a member of the Bridge Preservation Office staff, says, "The data collected is voluminous, and it's critical to the health of our bridges. Bridge inspection doesn't sound glamorous, but it is important work."[3]

EVERYDAY HAZARDS, EARTHQUAKES, AND LIGHTNING

Routine work varies on the bridge, including picking up motorists' litter and other debris on the roadway, and taking one-person elevators to inspect airplane navigation beacons on top of each tower. Working on a cable suspended hundreds of feet over cars and trucks as they whiz along the roadway below is mundane to the crew. Sometimes, they use the "spider," a one-man platform with a waist-high cage that uses a motor-driven winch to take a maintenance crewman up or down to a repair site.

Complacency that can come with routine is a challenge—when a crewmember clips a safety harness to a line and dangles hundreds of feet over the deck or the water, it is important to pay attention. On rainy days, the crew normally stays off the main cables and other heights, focusing on other chores, such as inspecting concrete at the anchorages. There also is a lot of shop work, including machining small parts and repairing equipment.

Danger comes with the territory. The tasks may appear simple to motorists who see the crewmembers on the span, but the job is rarely easy and always hazardous. Kip Wylie, Maintenance Lead Jon Moergen, and the rest of the crew routinely work at great heights and during rough weather. On any given day, they do not know what will come up. "That's the best part of the job," says Moergen. "It can also be the worst."

The story of 58-year-old Bill Clark is an example. Clark has been with WSDOT since 1993. In August 2004, Clark and another crewman were conducting a routine chore, greasing the connections on the hydraulic damping jacks at mid-span. Suddenly, a piece of lumber (an 8-foot-long 2-by-4) flew off the back of a passing truck at 55 mph, striking Clark in the head. His regulation hard hat saved his life that day, but he sustained a brain stem injury. Although Clark only stayed in the hospital overnight, he missed more than a year of work before returning to WSDOT in September 2005 and today remains plagued by periodic dizzy spells. The injury means that Clark will never again work on the Narrows span, or any other bridge.

One of the most memorable incidents for Wylie occurred during the winter of 1997 after a thunderstorm struck, with rain practically coming down sideways in sheets driven by high wind. Wylie received a call from the Washington State Patrol, informing him that the navigation beacon on the east tower was not working. Wylie went to the top of the tower, battling the freezing rain and buffeting gusts. What he found was surprising. Lightning had struck the beacon, popping the huge fixture's two bulbs, but there was no other damage.

An ice storm that same winter took out local electrical power for three days and left the entire bridge coated in two inches of ice. Since the elevators were inoperable, crewmembers had to climb 280 feet to the top of the towers using the hand ladders in the tower legs. Falling icicles prompted closure of the bridge for some 16 hours.

In February 2001, the Nisqually earthquake that struck the Puget Sound region proved a little too exciting for crewmembers. Jon Moergen vividly recalls that day. The bridge swayed more than he had ever seen before, first one way, then back the other. There was no damage when the shaking was over, except for a bent housing where one of the main cables connected into the east anchorage. Nonetheless, recalls Moergen, "It was scary." For someone who thinks nothing of walking up a main cable or dangling hundreds of feet over the chilly tides of the Narrows, this was saying a lot.[5]

Several big windstorms have blasted through the Narrows during the last five decades. One hit the bridge with winds up to 75 mph, and Wylie recalls another day when gusts of nearly 100 mph slammed into the span. A few times, officials have closed the bridge to avoid having cars blown into oncoming traffic lanes. The 1950 bridge, however, has never "galloped" like its 1940 predecessor.

WORKING TO KEEP TRAFFIC MOVING

Maintaining the flow of traffic remains a top priority when scheduling repairs and inspections. Most commuters seldom see or notice the maintenance crew as they are working. Routine repairs are scheduled for times with the least traffic, usually on weekends in the late-night and early-morning hours. Whenever possible, even emergency repairs are scheduled to minimize traffic disruptions.

And indeed, non-routine events do happen. Accidents frequently occur on the bridge—not just automobile fender benders, but incidents that cause damage to the span itself. For example, a truck hauling a submarine part to a U.S. Navy installation on the Peninsula tried to squeeze into just one traffic lane with its oversize load. The result—five long holes in the deck's pavement and a dent in one of the towers. Barges or boats often ram the piers—a barge hauling a 220-foot crane

LEGACY OF THE 1949 EARTHQUAKE

The east-tower cable saddle that plunged into the Narrows during the 1949 earthquake has suffered long-term effects from spending three days on the bottom of the Sound. Today, that cable saddle rusts more quickly than the others. "It seems likely," says Kip Wylie, "that the cast steel absorbed just enough salt water that paint doesn't bond to the surface as well as on the other saddles, so it corrodes faster."

struck the bridge's center span, causing some structural damage. Large vehicles crash into or scrape various parts of the structure, including a recent hit on one of the main cables that tore off part of the wrapping wire.

Crews work in the middle of the night and in all kinds of weather, even in howling wind and rainstorms. The public rarely notices—that's the way it's supposed to be.

Notes

1. Paul O. Anderson, "Workers Give Tacoma Bridge Bolt Overhaul," *Tacoma News Tribune,* September 26, 1954; Earl White, interviews, October 2003, November and May 2004, August and September 2005.
2. Kip Wylie, interviews, September and October 2003; Arvid Grant Associates, "Tacoma Narrows Bridge Condition, 1983," for the Washington State Department of Transportation; Arvid Grant Associates, "Tacoma Narrows Bridge Report, August 1991," for the Washington State Department of Transportation.
3. Sharan Linzy, interview, August 2005.
4. Jon Moergen, interviews, September 2003, May 2004.

CHAPTER FOURTEEN

PEOPLE OF THE 1950 NARROWS BRIDGE

MEN OF STEEL; NERVES OF STEEL

The tradesmen employed on the 1950 bridge, as with its 1940 predecessor, were union men who did the demanding physical labor of building the span. Also, as with the first Narrows Bridge, the professional engineers and managers of the Washington Toll Bridge Authority, Bethlehem Pacific Coast Steel, and John Roebling's Sons Company were non-union.

For the most part, roles were similar in both the 1940 and 1950 projects. The designers and engineers prepared the project's plans, specifications, and drawings. The "dimension control" group conducted surveys at the site, while the "inspection control" crew reviewed each stage of the construction work to insure accuracy and quality in the bridge. Ironworkers along with electricians, carpenters, laborers, and other tradesmen came from the ranks of the local population or were "boomers," who found their way here from other high-elevation construction jobs across the country.

As with their predecessors in 1940, the workmen who built the 1950 span were strong, hard-edged, quick-witted, and often colorful. "Bridge workers are a different breed of cat," said Earl "Whitey" White, a steel man on the 1950 bridge. They were a rough group,

and loved crude humor and practical jokes. The difficult, dangerous work bonded them. They did not always get along, but there was a sense of camaraderie.[1]

"Nerves of steel" is a cliché, but it best describes the steelworkers and other men that built the 1950 span. Daily they risked serious injury or death. Any of them would have told you, "You're not an ironworker unless you've been busted up somewhere along the line."

A division of labor was proscribed at the construction site. "Bridgemen" were welders, who positioned and welded steel, while "rivet men" heated, caught, and drove rivets. Apprentices, or "punks," assisted the bridgemen, fetching needed equipment and materials, while sometimes trying their hand with the tools and always observing and learning. Usually, an apprentice became a regular bridgeman after two or three years of experience. A first-level supervisor, called a "pusher," oversaw several groups of bridgemen and apprentices. All operated under the watchful eye of a "walkin' boss," who reported to the construction supervisor.

EARL "WHITEY" WHITE (B. 1921)

"Whitey" is a tough guy—and a survivor. He "worked iron" for 43 years on all kinds of structures before he retired in 1989. He grew up in Puyallup and settled in Tacoma. He and his wife, Penny, raised two sons on the modest income of a welder and heavy construction steelworker.

Earl White was born in 1921 in Wibaux, Montana. When he was three years old, his parents moved to Puyallup, where Earl graduated from high school in 1939. After a year in the Civilian Conservation Corps, young "Whitey" had enough money to take a

Harry Takahara (left) and Art Knoll (right) during the dangerous task of assembling the catwalks, September 1949.
WSDOT

Completing the west side deck, spring 1950. Front row (left to right): "Junior" Christiansen, Leroy Tober, and Ken Campbell. Back row (left to right): Glen "Whitey" Davis, Earl "Whitey" White, and Rip Sargeant.
Courtesy of Earl White

welding course, and by 1940 started working at the Tacoma shipyards. When World War II came along, he joined the U.S. Marine Corps at the age of 22 in 1943, and went off to the Pacific with the 5th Marine Division.

The spinning crew ready to start at the east end, October 28, 1949.
WSDOT

Back home after the war, Whitey again took up ironwork. From 1946 until he retired in 1989, he helped build scores of steel structures in the Tacoma area and around the Northwest, including high-rise office buildings, radar towers, dams, and, most memorable of all, the 1950 Tacoma Narrows Bridge.

"Working iron" is hard physical labor and injuries are common. Whitey narrowly missed death and had parts of his body mangled. Several times he had his "lunch box brought home" by a co-worker, meaning Whitey had gone to the hospital for an injury. One incident laid him up for 22 months.

At the Narrows from 1948 to 1950, Whitey worked on the towers, the cable spinning, and erecting the steel truss. The work was hard and conditions sometimes harsh. He saw a couple of friends die. With a somber, steady voice, he talks about the day that his buddy, Whitey Davis, "went all the way down," and "when he hit" the water "it sounded like an artillery piece went off." He saw another ironworker, Stuart Gale, go "in the hole" along with "40 ton of steel." He suffered through the severe winter of 1949–1950, when bitter wind, rain, and snow, and icy catwalks, made working conditions miserable for weeks.

Whitey routinely faced danger while helping to construct the span. He and the other bridgemen labored hundreds of feet over the Narrows without tying off their safety lines. They felt safer being unfettered. But working at great heights was not for everyone. When up high, Whitey noted, "I've [even] seen guys who've worked iron all their life freeze."

Steelworkers were "a different breed of cat," says Whitey. They were rough, and worked and lived hard. "And damn near every one of them would give you the shirt off his back." The same can be said of Whitey and he would take it as a compliment. Here is more of Whitey's story, in his own words:

"When they started the Narrows Bridge, I went out there and proceeded to help build the bridge for the next 29 months. I worked out of Local 114 Ironworkers Union; went to work for Bethlehem Steel. I think I earned $1.75 an hour. I started on the job when they were setting the tower base plates. I worked on compacting the cable, building the deck, and all the rest.

"We had a lot of boomers on the bridge. We brought in many gangs of riveters. Hot shot riveting gangs would come from 'Frisco and the East Coast. We had 'em from all over the world. Some would stay and some would get a paycheck or two and take off.

"That winter [1949–1950], the catwalk froze solid. Guys would take a few steps and their feet went right out from under them. When I got up to the top of the tower, I chipped ice off the cable saddle [during spinning]. You may not believe this, but that ice was 1⅛-inches thick. I suffered more on the Narrows Bridge than working on radar towers in the winter in Cutbank and Havre, Montana. It was just so cold and wet. It was bitter

Earl White, second from right, and fellow ironworkers in the winter of 1949–50, one of the coldest on record.
Courtesy of Earl White

cold. Snow was drifting, wind was blowing. It was miserable. One time during the night, the wind kept picking up. It was over 100 miles an hour. That was the scariest night I ever put in in my life.

"When we'd start up that cable, I could almost put a paint mark where I took the next step when that wind would come around that point and hit you right square in the face…'smack.' All of a sudden, you leaned forward, because from then on you were going to lean forward. Then, sometimes I'd go up that catwalk in the morning. It would be socked-in solid; you couldn't see your hand in front of your face. You know how fog sets in. You'd get up above that 500-foot level, and all of a sudden your head come up above those clouds and you felt like you're in heaven. You'd see the mountain tops sticking out, just a little bit of them. It was like being in an airplane up above the cloud level. It was an eerie feeling sometimes.

"Working on the bridge, you'd see pods of killer whales come through. Then, there comes a ball of herring 200 feet across—looked like a big basketball underneath the water. Then you'd see the salmon, swimming through, hitting them with their tails, then they'd come back and eat 'em. Once in a while, you'd see a seal after the salmon. Then, the boats would come. When the boats saw the seagulls, they'd come up to get a supply of herring for fishing.

"The wives had a pretty hard time too. Like that day Whitey Davis fell off. We just had radios in those days. Well, I was known as 'Whitey' too, and I had light hair then too. One gal called my wife and told her, 'Whitey fell off the bridge.' We closed down the bridge and headed to town and held a wake. She found out in the meantime it wasn't me. But, for a long while all she knew was, 'Whitey went into the hole.' Then, this gal calls up my wife and says it wasn't Earl. I got home about midnight. I wasn't very popular. She'd had my lunch bucket brought home three or four times. You get busted up quite a bit.

"I've seen guys who've worked iron all their life freeze. On any given day, just freeze. In those days we never tied off, like now. We had our ropes on—you had to wear this. But, we never tied off because about the time you tie-off you have to make a quick move, something's gonna happen. We just didn't feel safe.

"Bridge workers are a different breed of cat, I guess that's how I'd describe 'em. I had some wonderful friends in the ironworkers, but there's only about two of 'em I've ever brought into my home. They're a pretty rough bunch. They worked hard, they lived hard. And, damn near every one of them would give you the shirt off his back.

"One Christmas I'd been out of work. A load of iron bars knocked me, and I come down and smashed my elbow all to hell. Still, I can't straighten it out. That Christmas things were tough, real tough. Come Christmas, three of the ironworkers come up to the door, and they had everything you could think of for a Christmas dinner, turkey, wine, cranberries. That's the kind of guys they were."[2]

JOE GOTCHY (1903–1994)

Quite a few men helped build both the 1940 and 1950 bridges, but one who has left a special mark in history is Joe Gotchy. At the age of 87, Gotchy wrote a personal account of his experiences. His book, *Bridging the Narrows* (published by the Gig Harbor Peninsula Historical Society, 1990), is an important source of information about the construction of the two bridges and the people who worked on them.

Gotchy handled various jobs on the two spans. He supervised concrete mixing and operated cranes during pier construction. On the 1950 bridge, Gotchy worked as an operating engineer helping to build the towers, then participated in erecting the stiffening truss with the west-side crews from the Gig Harbor end toward mid-span.

Joe Gotchy loved bridges. Born on February 3, 1903, in Bothell, Gotchy grew up in

Joe Gotchy, ca. 1990.
Gig Harbor Peninsula Historical Society

rural Thurston County. He worked on bridges all over Washington. Though a high school dropout, he read avidly and became a life-long learner. In his career, Gotchy was an operating engineer, ironworker, and a lumberjack. He cared deeply about his family, fellow workers, community, and the environment. Gotchy and his wife resided in Gig Harbor for 23 years, before he passed away on December 3, 1994.

In 1992, 89-year-old Gotchy helped gather signatures to petition the State to provide funding for a new Narrows Bridge (2007) that would ease traffic congestion. Joe Gotchy is one of the very few who have contributed directly and indirectly to all three Tacoma Narrows bridges.[3]

BILL MATHENY (B. 1915)

One-time ironworker Bill Matheny claims a unique distinction. Now 90-years old, he is the only man to get a union hall dispatch to work on all three Tacoma Narrows bridges.

Matheny was born on December 24, 1915, and raised in Tacoma. His grandfather came west from Kansas on a wagon train and homesteaded here in 1855. His mother's father, he notes with pride, was a Tacoma policeman. Matheny grew up on the south side of Tacoma in what he calls a "tenement house." He graduated from Lincoln High School in 1934, but the Great Depression made it difficult to find work. He took a job in a local sawmill, then spent a couple of years as a logger in Oregon. When news got around in mid-1938 that there would be a bridge built at the Tacoma Narrows, Matheny wanted to get on a crew.

"I tried to get into the Ironworkers Union," he recalls, "but unless you had a father or grandfather who was a member, you didn't have a chance." Matheny is a man of determination. He went down to the union hall almost every day and hung around. There were other lumbering jobs in the meantime, but finally in January 1939, he was sent to the Narrows to work on the cable-spinning phase.

Crewmembers observe the grooved trough of the cable saddle; the guide wire for spinning is in place.
WSDOT

First, he served as a "permit man"—an apprentice ironworker at $1.00-per-hour wage. In 1939, that was good money. The "permit men" were not in the union, but received a special permit to work on the bridge. "They needed a lot of men for the cable spinning," explains Matheny, who participated in one of the three round-the-clock shifts. His post was at one of the spinning stations, where he guided each wire brought by the spinning wheel into a piece of channel iron. Once the wires were formed into a strand, a crane lifted the strand onto the cable saddle. His other apprentice duties over the next 18 months included helping with the cable compacting.

Break time for a rivet gang assembling the suspended structure, spring 1950. The man at far
right wears a bucker-up's hat.
Gig Harbor Peninsula Historical Society, NB-164

"Then," smiles Matheny, "I worked what
they called 'punkin' rivets.'" As a "punk," he
hauled rivets to the rivet gangs and bolts to
ironworkers. Matheny remembers the boomers as "hard drinking, hard working, husky"
men. They had traveled all over the country,
it seemed to Matheny. Many had come north
from San Francisco. When the Narrows
Bridge neared completion in the spring of
1940, Matheny was tying wire around rebar
in the roadbed, prior to concrete pouring.
"That bridge was really bouncing around,"
he says. "I remember a guy saying, 'Hey, this
thing ain't gonna last.' He was right, as it
turned out."

At the grand opening on July 1, 1940,
Bill and a handful of other young apprentice
ironworkers stood on a hill nearby, watching the festivities. "We felt a touch of pride,"
he emphasizes. "Hey, we helped build that
bridge. To my knowledge, I don't think there's
any of those guys left alive."

After work at the Narrows Bridge ended,
some men found jobs at the Bremerton shipyards. Matheny had finally received his long-
sought union card, and wanted to stay with
ironwork. He found it at nearby Sussman
Steel. One day somebody called out, "Hey, the
bridge fell down." Matheny dropped his tools
and went out to the Narrows. "I had to go see
it," he says.

After the outbreak of World War II on
December 7, 1941, many of the young men
who had worked on the Narrows Bridge went
into the armed services, including Matheny.
He served as a bombardier on a B-29 bomber
in the Pacific. In August 1945, he was stationed on Tinian Island when the "Enola
Gay" took off from a field on the south end of
the island to drop an atomic bomb on Hiro-
shima, Japan. At the same time, Matheny's
group took off from the island's north field
on a different mission. Upon returning, he
learned it had been an historic day—everyone
was talking about the "atomic bomb" dropped
by a B-29 stationed on the south end of
Tinian.

After the war, Matheny went back to construction work. In 1948, he hired on with
the new Tacoma Narrows Bridge project at
the cable-wire reeling plant on Tacoma's tide
flats. There, big rolls of wire arriving by railroad in gondola cars were lifted out with a
forklift and taken into the plant, where they
were unwrapped. Matheny ran a splicing
machine, connecting two ends of wire before
they were put on big reels and then sent out
to the bridge for the cable spinning. Later,
Matheny worked on the bridge itself, tight-
ening bolts on the cable bands. It was a two-
man job, using a six-foot wrench to properly
secure the bolts.

With the end of the Narrows project,
Matheny worked on other bridges, either
connecting iron or on riveting gangs. He
also helped put up buildings in Tacoma and
the south Sound area. Then fate changed
his career. One day in Aberdeen, a load of
iron fell, mangling his ankle. That ended

his iron working career. After he recovered, he was employed by Prudential Insurance until retirement. Bill and wife Betty Ann celebrated their 60th wedding anniversary in 2002. The couple has a daughter and two sons, and numerous grandchildren and great grandchildren.

Bill Matheny remains an ironworker at heart. "I enjoyed iron work," he grins, "every minute of it." He points with pride to his "Honorary Dispatch" to work on the new 2007 Narrows Bridge. "I'm the only one alive who has had a union dispatch to work on all three Narrows bridges."[4]

ARNIE COLBY (B. 1916)

The "dimension control" men who performed the survey work largely have been ignored by history. The work was rarely dramatic. Arnie Colby knows that and does not mind. Or at least he has become accustomed to it, as have most civil engineers. The story of this veteran of the 1950 bridge project is remarkable, espe-

cially because it shines light on a little-known facet of bridge building.

Colby was born in 1916 and grew up in the countryside near Auburn, Washington. By 1934, he had graduated from high school and began engineering studies at the University of Washington. He had a keen sense for details and wanted to pursue a career that would keep him outdoors. Although he did not graduate, he took enough classes to become a certified professional engineer. In the midst of the Great Depression, he hoped the education would give him a chance for work.

He married and found a succession of jobs on major public works projects in the Northwest. He performed survey duties for the U.S. Bureau of Public Roads (now the Federal Highway Administration), a roadway to Mt. Rainier National Park, power-line corridors for the Bonneville Power Administration, and during construction of Grand Coulee Dam. In May 1946, Colby took a job with the Washington State Highway Department, where he remained until retirement 30 years later.

At the age of 32, when he heard that the Narrows replacement bridge project was starting, Colby was able to transfer to the WTBA and worked on the survey team for the next two years. The survey crews set elevation levels for the remodeled piers, relaying information to the supervisors, such as, "the steel plates for the tower base plates are two-thousandths of

Steelworkers assembling a truss, one of the most hazardous phases of construction.
Courtesy of Earl White

Driving the last rivet, 1950.
WSDOT

an inch too high." During tower construction, Colby climbed to the top, and while holding on with one hand, leaned over sideways with a target in his hand so that the transit-man could "shoot" an elevation.

"It's all precision," explains Colby. With transits, levels, targets, and "chains" (metal measuring tapes), the dimension control crew set the control points for each step of construction. Directing the placement of steel beams and board forms for concrete were all in a day's work; they shared the same hazards and heartbreaks, heat and rain, ice and wind, success and pride as the carpenters, ironworkers, welders, laborers, and others.

"The final product is the culmination of a long series of little steps," explains Colby. "If each one is not exact, the thing won't fit together. When one crew is building the steel deck a half way across the Narrows from the east, and another one is doing the same thing from the west, and it's a mile across, it's not an accident that it came together without a hitch." The bridge's suspended structure did fit together, perfectly.

For two years after the Narrows job ended in 1950, Colby helped survey for the cross-Sound bridge project. In 1952, he returned to the State Highway Department. Since his retirement in 1976, Colby has continued to live in Auburn. He and his wife are proud of their grandchildren, great grandchildren, and even great-great grandchildren.[5]

KIP WYLIE, MAINTENANCE SUPERVISOR

The current head of maintenance operations, Kip Wylie, is a quiet man with a passion for windsurfing and restoring his 110-year-old historic home. He also is passionate about the span he is responsible for maintaining. To walk around the Narrows Bridge with him is to plant one foot in the past and the other in the future. Wylie provides a wealth of facts and details about routine maintenance and bridge history, lacing his commentary with odd anecdotes. At the same time, he points out current and future problems, and how

they can be managed to keep the bridge operational for the public.

Wylie has been assigned to the Narrows Bridge since 1983. Initially, he was an experienced painter, but by 1988 became the lead technician, or "line boss." However, his duties remained about the same—it still was a working position. As maintenance supervisor, Wylie manages a crew of five people—three painters and two ironworkers. Constant replacement and repair work is required on the Narrows Bridge, which now is well over a half-century old. Wylie not only schedules the crew's work, but also manages the maintenance budget—$1.7 million for a two-year period. He also handles an assortment of safety issues in cooperation with local fire and police departments, and, on any given day, you may find him repairing phone lines or ordering supplies.

Wylie has seen some terrific windstorms strike the bridge. The highest wind he recalls was clocked at 100 mph. As for the span's movement, it is hardly noticeable. It might seem that practically "nothing happens," says Wylie. That is because the hydraulic damping jacks and the stiffening truss below the roadway keep all movements in the span very slow and virtually imperceptible. "Even in a high wind," he notes, "I'll sight down the deck and see the bridge is moving only slightly." The label once applied to the bridge, "Sturdy Gertie," has not become a widespread household word, but, says Wylie, "It's justified."

The Narrows Bridge is almost his second home. "I live with it all the time," says Wylie with a smile. "I know its sounds, creaks, groans, and movements. I feel like this bridge is a member of my family."[6]

JON MOERGEN, MAINTENANCE LEAD

Today's maintenance lead on the Narrows Bridge is Jon Moergen. This genial, husky, middle-aged man with a prominent moustache is called "Moergen" by his friends. He spends leisure time with his dog, two cats, and a 1930s-era powerboat that he is restoring.

Jon Moergen, maintenance lead, 2003.
Richard S. Hobbs

On any given day, he might be repairing a $2,000 arc welder in the bridge's shop or walking a main suspension cable to check maintenance needs. Modest and matter-of-fact in his approach to the daily routines, Moergen is proud of the almost 20 years he has worked on the span. "It's good maintenance that has kept the Narrows Bridge functioning and in good shape for over 50 years—so you *don't* notice problems."

Occasional visitors ask Moergen questions that range from the predictable to the bizarre. Most often, he hears, "The bridge moves? I didn't realize the bridge moves." That brings a smile, and a patient reply that wins friends. "People look at the bridge as a static structure. They just can't seem to grasp that it is a constantly moving machine. Once they get the idea, they begin to think about the stresses the bridge must withstand."

Like Kip Wylie, Moergen also retains a hefty measure of historical facts, odd tales, and personal experiences. Some of his daily routine is hair-raising to non-bridge folk. And the tough, dangerous work has taken its toll;

Moergen has had his share of injuries and narrow brushes with death.

Ask Moergen what really keeps him coming back, day-in, day-out, and he will say he likes working outside, especially on warm summer days. But he adds something unexpected with a dead-serious stare: "The most invigorating thing about this job is the *unknown*. You never know what to expect when you come to work each day. It's always a challenge. Some days, that's also the worst part of the job."[7]

SHARAN LINZY, WSDOT BRIDGE PRESERVATION OFFICE

Sharan Linzy of the Bridge Preservation Office is one of the few women doing bridge inspections. She is unusual in the engineering world for other reasons as well. Linzy is afraid of roller coasters and climbing high ladders up to rooftops. Yet she is comfortable at great heights in the lift bucket of a UBIT (Under Bridge Inspection Truck), or walking on a main suspension cable. "Oddly enough," she

Sharan Linzy of the WSDOT Bridge Preservation Office sits comfortably on a cable saddle, 2005.
Erick Sniezak

muses, "I've been fascinated with bridges since I was a little girl."

Now in her early forties, Sharan Linzy's path to an engineering career has been a zigzag. She was born at Andrews Air Force Base in Maryland. When Linzy was four years old, her parents moved to California, and by the time she was 12 they settled on a farm near Olympia. She grew up enjoying 4-H Club and Future Farmers of America activities, showing horses, dairy cows, and rabbits. She started working for the State over a decade ago, with most of that time at WSDOT.

In 2001, Linzy decided she had waited long enough to pursue her dream job. At first, she set her sights on becoming a bridge inspector. Before long, she also decided she wanted to be an engineer. She impressed Harvey Coffman, head of the WSDOT Bridge Preservation Office, with her talent and determination just at the right time. A bridge technician position opened, and Linzy was on her way. She also went back to college. Over the last several years, in her "spare time," she has been taking night classes, earning high grades toward a bachelor of science degree in civil and structural engineering at St. Martin's College. "Homework is my hobby," she smiles.

The Bridge Preservation Office experience is a great education in itself. Now a "bridge engineer," Linzy handles multiple jobs around the office, including responding to public records requests and maintaining the State's bridge inventory database that WSDOT sends to the Federal Highway Administration each year. She tracks bridges from the time they are designed, through construction, and to completion. Linzy is qualified as a co-inspector and accompanies a lead inspector regularly during the summers to examine bridges around the region.

As an assistant to a lead bridge inspector, Linzy takes measurements and photographs, compiles notes on the condition of various bridge parts, and assists in writing reports. She looks for cracks in concrete, rust and other corrosion on exposed steel, defective bolts, and other signs of structural deterioration. By performing these tasks and providing a valuable alternate view of bridge site conditions, she helps the lead inspector work more accurately and efficiently.

In the future, when she has a little leisure time to enjoy traveling with her husband, she plans to visit Australia. More accurately, she wants to climb a bridge in Australia, the great steel arch Sydney Harbour Bridge. "They let the public walk all the way over the arch, across the harbor, and it only takes three and a half hours," she says enthusiastically.[8]

Notes

1. Earl White, interviews, October 2003, November and May 2004, August and September 2005.
2. *Ibid.*
3. Joe Gotchy, *Bridging the Narrows* (1990); Anthony Albert, "Joseph Gotchy Remembered as Man of 2 Bridges," *Tacoma News Tribune*, December 14, 1994; Bart Ripp, "'Bridging the Narrows,' Joe Gotchy's Book Spins Tales of Men and Steel," *Tacoma News Tribune*, July 3, 1990.
4. Bill Matheny, interview, September 2005.
5. Arnie Colby, interview, September 2005.
6. Kip Wylie, interviews, September and October 2003, August 2005.
7. Jon Moergen, interviews, September 2003, May 2004.
8. Sharan Linzy, interviews, August and September, 2005.

1940 Bridge

Galloping Gertie shortly after opening day.
WSDOT

Matchbook cover.
Courtesy of Gig Harbor Peninsula Historical Society

Souvenir program for the Tacoma Narrows Bridge opening, July 1, 1940.
Washington State Archives

1950 Bridge

Souvenir booklet for the opening celebration, October 14, 1950.
WSDOT

Looking west across the Narrows toward the Olympic Mountains.
WSDOT

Adjusting bolts on a suspender cable band, nearly 300 feet above traffic.
WSDOT

A cable saddle at the top of the 1950 bridge's eastern tower (2005 photo).
Sharan Linzy

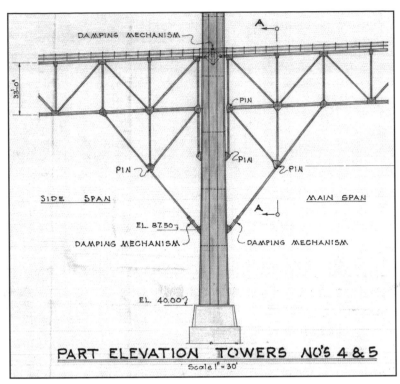

Part elevation of towers showing damping mechanisms,
by Dexter Smith, December 5, 1945.
WSDOT

Hydraulic dampers connect the suspended structure to Tower #5 (2005 photo).
Sharan Linzy

Eastbound traffic slowing to a crawl, increasingly common in
the 1990s.
WSDOT

The maintenance crew often works under the roadway unobserved by motorists.
WSDOT

During an inspection in 2000, a main suspension cable is unwrapped and wedges driven to separate the wires, so the interior can be examined for moisture, rust, and deterioration.
WSDOT

Following the 2000 inspection, Kip Wylie straddles a cable while positioning the cable compactor.
WSDOT

Kip Wylie (front) and visiting high-steel inspector Dan Weber observing workers from a convenient vantagepoint.
Richard S. Hobbs

Earl "Whitey" White (right), near the end of his last cable walk, converses with Kip Wylie at a cable saddle, August 23, 2005.
Richard S. Hobbs

Tower view.
WSDOT

A commemorative plate.
Courtesy of Gig Harbor Peninsula Historical Society

Seattle artist Sarah Clementson's popular Christmas card.
Sarah Clementson, seattlewatercolors.com

In 1949, bridge engineers made a Christmas card on blueprint paper.
Courtesy of Arnie Colby

2007 Bridge Construction

Early in the tower construction phase.
WSDOT

A new tower begins to rise on the south side of the 1950 span.
Sharan Linzy

Sharan Linzy walks the cable on a routine inpsection. In the distance, the new east tower nears completion.
Eric Sniezak

A construction crew celebrates tower completion.
WSDOT

Bright lights along the cable catwalks illuminate the construction site at night.
WSDOT

Roadway deck is lifted into place on the east side.
WSDOT

Sections of roadway deck are lifted from a barge to be suspended from the cables.
WSDOT

Decking is assembled on the west side of the new bridge.
WSDOT

CHAPTER FIFTEEN

STYLE FOR A NEW SPAN

> "Mere size and proportion are not the outstanding merit of a bridge; a bridge should be handed down to posterity as a truly monumental structure which will cast credit on the aesthetic sense of present generations."—*Othmar H. Ammann, 1954*

ARTISTRY

The State's engineers were keenly aware of Galloping Gertie's problems, and the spectacular failure loomed large in their efforts to plan a new span. But the design of the 1950 Tacoma Narrows Bridge was more than a simple reaction to failure. The artistic spirit embodied in the bridge differed markedly from that of its predecessor.

After World War II, the practical considerations and aesthetic trends that shaped suspension bridge design had changed. There were different program requirements and different personalities involved in planning the 1950 bridge. Standardization and pre-fabricated parts played an increasingly important role in construction methods. Rational and technologically sophisticated designs proliferated for bridges, buildings, and structures of all kinds. A host of new currents in the artistic, social, and economic life of the country also were part of the working environment for the designers.

Galloping Gertie had been widely described in "feminine" terms. At the time, even engineers used words like "elegant" when noting "her" slender, sleek profile. Tellingly, the new bridge, hailed for its strength and solidity, became known as "Sturdy Gertie." In the eyes of many, perhaps, it represented more "masculine" qualities. Regardless, there is artistry in the bridge.

BEAUTY AND STRENGTH

When completed in October 1950, the new Narrows Bridge was the most advanced suspension span of its day. Some writers, however, later criticized the design effort as "too cautious," while others even called it "ultra conservative." They characterized it as "over-

1950 Narrows Bridge tower. *WSDOT*

engineered"—i.e., it was much stronger than it needed to be in regard to the wind and traffic parameters at the Narrows.

Indeed, principal consulting engineer Charles Andrew and design engineer Dexter Smith intentionally adopted a design that would have the least possible chance of

Paying a toll in the early 1960s.
WSDOT

towers of the Narrows and the San Francisco-Oakland Bay bridges, it is noteworthy that Charles Andrew led the design and construction for both spans.

A remnant section from Galloping Gertie is often overlooked as one of the fascinating features of the 1950 bridge. The west-side span and its support (Tower #3) were undamaged when the 1940 span fell into the Narrows. They remain today, along with Eldridge's original two piers in the channel, as integral parts of the 1950 Narrows Bridge.

Bridge Singing

All suspension bridges can "sing"—the main suspension and suspender cables vibrate, much like guitar strings, and at different tones. Normally, these vibrations are inaudible, but nevertheless they occur. The Narrows Bridge's "singing" has been decoded as part of a broad effort to record and replicate these unique sounds for the public. Jodi Rose heads an international project (www.singingbridges.net) that captures and amplifies this "singing" as a "sonic sculpture." Rose enthusiastically proclaims, "Sounding the harmonic frequency within the unheard vibrations of the cables will release the voice and liberate the spirit of each bridge."[1]

In more a more traditional musical aspect, it appears that performers have not written any songs about the 1950 bridge,[2] but they did for its 1940 predecessor. Galloping Gertie was celebrated in at least two musical compositions. At the opening day ceremonies on July 1, 1940, bands played a special march for the occasion, composed by Gig Harbor's postmaster. Another tune was more prominent—"Ballad of Galloping Gertie"—recorded and played on Northwest radio stations after Gertie's collapse. It seems to have since disappeared, except in the memories of some local citizens who clearly recall hearing the song on the radio. Meanwhile, music buffs continue to search for a copy of the recording or the sheet music.[3]

repeating Galloping Gertie's problems. For bracing in the tower legs, they chose the structural form with the greatest strength—the "X." This cross-bracing pattern very effectively resists forces that pull or twist. An "X" also presents a dramatic expression conveying strength, power, and solidity. Aesthetically, it is the dominant design element in the 1950 bridge.

Each tower features three large panels of double-"X" cross-braces between the tower legs above the deck, and three single "X" cross-braces below the deck. The huge "X" braces below deck measure approximately 60-feet wide and 45-feet high. The cross-bracing pattern displays a sense of balance and proportion that is enhanced by the straight, vertical lines of the tower legs. The towers, connected at the deck by the 33-foot-deep "W" of the Warren truss, exhibit artistry and technical strength of unprecedented magnitude.

The massive "X" had been prominent in preceding spans. The 1937 Golden Gate Bridge, which influenced the design of the upper tower bracing of the 1940 Narrows Bridge, has two prominent "X" braces below the roadway. Other examples are found in such celebrated spans as the 1931 George Washington Bridge and the 1936 San Francisco-Oakland Bay Bridge. In comparing the

George Washington Bridge (1931).

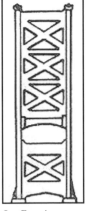

San Francisco-Oakland Bay Bridge (1936).

View of the bridge in September 1950.
WSDOT

View from the west shore showing the 1940 Tower #3 (foreground) that was incorporated into the new 1950 bridge.
WSDOT

The Bridge and Community Art

Residents throughout the Puget Sound region heralded the opening of the new Narrows Bridge in numerous ways. Besides customary parades, public speeches, and newspaper specials, the new bridge inspired artistic efforts in the local community. For example, Tacoma firemen made a display model of the bridge for the annual Chrysanthemum Society flower show in 1950. Over the years, the Narrows Bridge has been featured in various commemorative objects and collectibles—everything from ceramic plates to T-shirts and even Christmas cards.[4]

Today, during Gig Harbor's summertime street fair, artistic photographs of the span appear almost weekly. The bridge is commemorated in classroom projects, too. Recently, a collaborative effort between the students and staff at the St. Nicholas School in Gig Harbor and the new Tacoma Narrows Bridge Project produced some unusual watercolors for an online exhibit. The sixth-grade students based their inspiration on photos of the 1950 bridge as well as the ongoing construction of the 2007 span. Each painting represented one view, one moment, and one day in time for the new construction effort and the 1950 bridge.

Doug McArthur was one commuter who never tired of the Narrows Bridge. For years, the Tacoma native crossed the span twice a day to and from his job near Gig Harbor, where he served as marketing director for Canterwood, a newly developed residential community. One evening in 1989, as he drove homeward, inspiration moved him to jot down a poem, which he titled, "Ah, the Beauty of the Bridge."

McArthur formerly had served as athletic director at the University of Puget Sound. These days, he stays busy writing about Tacoma sports heroes. For nearly a half-century, he and his wife have viewed the bridge when taking the family dogs on nightly walks in their east-shore neighborhood at the Narrows. For Doug McArthur, and many others, the Tacoma Narrows Bridge remains an object of beauty.

Ah, the Beauty of the Bridge

Like twinkling strands of Christmas lights stretched along a roof
    Where beacons touch the darkened nights, with monumental proof
That frigid depths far below
    Can neither stop nor even slow
Man's uncompromising quest
    To link his days of work with rest.
Ah, the beauty of the bridge!
    In the morning standing tall, as the sun begins to call
The mountain looms at break of day, a welcome sight along the way
    Reminding you, driving there
To look around you, everywhere
    To realize what nature's done
For man to span to reach someone
    Ah, the beauty of the bridge!
When day concludes and homeward bound, you reflect on what you've found
    The Narrows edge is a treasure, serving man endless pleasure
Across that channel lies a world
    That causes all your cares unfurled
To vanish like the sunset skies
    Which stretch before your very eyes.
Ah, the beauty of the bridge!
    Looking back you realize how symbolic there it lies
Always ready, ever sturdy, far beyond a galloping "Gertie"
    Like a guardian day and night
Straight and tall, to the height
    Like a chain never broken
Like the words which go unspoken.
    Ah, the beauty of the bridge.

—Doug McArthur, 1989

Notes

1. Quoted from the project's web site, www.singingbridges.net/about/.

2. A 1962 folksong, "Steel Men," commemorates steelworkers and two different bridge collapses, but not Galloping Gertie or the men who constructed the 1950 bridge. Written by David Martins and sung by Jimmie Dean, it is part of a working-songs collection, maintained by Folklore Heritage in the Pacific Northwest. See: collections.ic.gc.ca/folklore/worksong/steel2.htm.

3. Two documented sources mention the tune, or perhaps two tunes. Clark Eldridge's unpublished "Autobiography" refers to the 1940 bridge being celebrated in a song called "The Galloping Gertie." Also, a description of a "Ballad of Galloping Gertie," written by Jim Moore, appears on a photocopied page from an unidentified publication in the Gig Harbor Peninsula Historical Society collection. The 1950 bridge's 50th anniversary also moved local poet Bette Dawson to pen a rhyme about the 1940 span, taking its title from the song,

"The Ballad of Galloping Gertie." A copy of Dawson's work is at the Gig Harbor Peninsula Historical Society Museum.

4. The first two Tacoma Narrows bridges continue to inspire and influence people in many ways. Tubby is the subject of at least one web site. At WSDOT's 2007 Tacoma Narrows Bridge Project building in Gig Harbor, engineer Dennis Engle (an avid collector of TNB memorabilia) has named the office's meeting area the "Tubby Memorial Conference Room." The recent works of composer and "sound designer" David Kristian, known for his unusual electronic "soundscapes," include an album titled "Tacoma Narrows Bridge." The popular Canadian rock band, The Tragically Hip, from Kingston, Ontario, uses the collapse of the 1940 span for lyrics in the song "Vaccination Scar." An image of the 1950 bridge appears on the cover of David LaRosa's recent non-fiction book, *Tacoma Confidential,* about the 2003 murder-suicide involving Police Chief David Brame and his wife, Crystal.

A computer enhanced view, looking west, showing how the completed 2007 bridge (left) will appear next to the refurbished 1950 bridge (right).

Tacoma Narrows Constructors

A Bridge to the Future, 1993–2007

If you stand on a bluff above the Narrows today you will witness history in the making. The newest Tacoma Narrows Bridge is the longest suspension span erected in the United States since 1964. Scheduled for completion in 2007, the $615 million span has been more than a decade in the making. In fact, the first public suggestion that a second bridge was needed at the Narrows dates to 1965, when the 1950 bridge was only 15 years old. Nonetheless, it took another two decades, and a lot more traffic snarls, to launch serious efforts to build a new span.

By the beginning of the 1990s, traffic congestion on the State Route 16 (SR 16) corridor was a major headache for commuters and others. Rush hour traffic had grown substantially beyond the roadway's capacity. Many days saw 85,000 to 90,000 vehicles drive over the bridge. By the year 2000, the count topped 100,000 on peak days. Everyone wanted to improve the safe movement of people and freight through the SR 16 corridor.

The traffic lanes of the 1950 Narrows Bridge were designed for the cars of that era. More than five decades later, the lanes are narrow by contemporary standards. The open grating separating the lanes tends to impede traffic flow. Also, there are no shoulders, nor a physical separation of opposing traffic movement. The bridge towers are located right at the outside lane edges. The 1950 bridge's physical constraints, and the over-capacity vehicle usage, have contributed to the significant congestion problems on SR 16. At the same time, normal WSDOT revenues have not been sufficient to keep up with the state's new transportation requirements. Consequently, new innovative methods of developing and financing projects have emerged.

Milestones to Groundbreaking

The first milestone came in 1993. That year, the State Legislature passed the Public Private Initiatives Act (PPI). This legislation authorized the Secretary of Transportation to solicit and select up to six demonstration project proposals. By January of the following year, the WSDOT advertised a request for proposals for transportation alternatives in the SR 16 corridor that would be funded by public-private partnerships.

At this point, the electorate became more directly involved. In its 1995 session, the State Legislature amended the PPI act, which required a public vote for the Narrows project. In preparing for the advisory vote, WSDOT conducted a comprehensive analysis of traffic patterns and economic effects to define the geographic boundary of the area that would be affected by tolls on a new bridge. Once this was defined, a citizens advisory committee was formed to advise WSDOT on all aspects of conducting the advisory election.

Meanwhile, as the preparations were underway for the ballot issue, a 1997 Major Investment Study (MIS) identified 22 potential multi-modal alternatives to improve transportation efficiency along the SR 16 corridor, including a toll bridge that would parallel the current Narrows Bridge. Also during this period, the alternative of adding a second deck to the existing bridge was studied in the MIS process. The 1950 bridge was not designed for widening at deck level, though a second deck on a strengthened set of bottom truss laterals could be added to increase capacity. However, the towers, footings, and anchorages would require major modifications to accommodate a second set of main cables for carrying the weight of a second deck. The financing of these modifications was quite close to the costs of a new bridge. Furthermore, a remodeled 1950 structure would not achieve the same safety benefits nor flexibilities that a new span would provide. The second deck option ultimately was not selected as a preferred alternative.

The public's advisory vote on the project was included in the November 1998 general election ballot. Voters were asked, "Should the Tacoma Narrows Bridge be modified and a parallel bridge constructed, financed by tolls on bridge traffic and operated as a public-private partnership?" A majority of the public answered "Yes." The 1999 legislative session next approved a $50 million appropriation as a contribution. Meanwhile, WSDOT selected 20-year veteran state engineer Linea Laird to head the project. Laird, an Alaskan native and veteran of many civil projects (including a stint on the Alaskan pipeline), had the task of bringing the mammoth effort together. Under her leadership, a new bridge at the Tacoma Narrows started to move rapidly toward becoming a reality.

The State initiated an elaborate process to identify and define preferred solutions for improving the SR 16 corridor. Environmental impact studies were launched and applications were made to secure the appropriate permits from government agencies. The Final Environmental Impact Statement (FEIS) was issued by the Federal Highway Administration in 2001. At that point, more than 20 permits were obtained from federal, state, and local agencies.

The FEIS, available to the public at local libraries, reviewed a host of considerations and impacts anticipated in the construction of the new bridge—noise, air quality, geology, soils, land use, zoning, recreation, wildlife, fisheries, biota, social and economic concerns, public services and utilities, visual quality, hazardous materials, and historic, cultural, and archaeological resources. Some of the more important environmental improvements addressed included the relocation of the exist-ing Living War Memorial Park, providing new and improved storm water collection facilities, protecting and reestablishing wetlands, and developing and monitoring a new deep-water environment for marine life.

Now the stage was set for moving the huge project to completion. In 2002, with the State Legislature's earmarked appropriation and voter approval of the bond issue, WSDOT assumed management responsibility for construction and operations. On July 16 of that same year, WSDOT executed a fixed-price, design-build, agreement with Tacoma Narrows Constructors, a joint venture of Bechtel Infrastructure and Kiewit Pacific Company. By September 18, the first bond sales opened for the initial funding.

On October 5, 2002, Washington State Governor Gary Locke led a host of government officials and other speakers at a public ground breaking ceremony. Design for the new bridge began immediately. The entire construction effort, including improvements to the 1950 Narrows Bridge, was on a 5½ year schedule.

SHAPE OF THE FUTURE

In May 2003, aerodynamic testing of the new bridge design was conducted in Ottawa, Canada, by Rowan, Williams, Davies & Irwin, Inc. in cooperation with Canada's National Research Council, Institute for Aerospace Research. For the first time ever, two complete models of suspension spans—the 1950 and the 2007 bridges—were tested side by side in a wind tunnel. The goal was to identify if the new structure, due to its proximity to the 1950 span, might cause significant wind interference that could damage either or both bridges. Because the models reacted in a wind tunnel just like real bridges, the research revealed that the "twin bridges" would co-exist safely. The models held steady with no torsional flutter in a 160-mph wind.[1]

Construction of the new parallel suspension bridge began south of the 1950 span.

Aerodynamic testing of the 1950 and 2007 bridges, the first ever wind tunnel trials on parallel suspension bridge models. Rowan, Williams, Davies & Irwin, Inc. conducted the tests in cooperation with Canada's National Research Council, Institute for Aerospace Research. Left to right: Marco Accardo, Instrumentation Specialist; Stoyan Stoyanoff, Aerodynamic Specialist; and Brian Dunn, Assistant Project Manager.
RWDI of Guelph, Ontario, Canada

Unlike the previous Narrows bridges, the 2007 structure did not have a person assigned as principal designer. Its design was a collaborative team effort. Tacoma Narrows Constructors selected the joint venture firm Parsons Transportation Group (PTG) and HNTB, Inc., to design the bridge and roadway approaches. PTG is the modern-day incarnation of the famous New York bridge building company founded by David Steinman.

Tim Moore was assigned as WSDOT's senior structural engineer to handle the technical design review, ensuring that the bridge would meet all state and federal codes. Today, Moore notes that he has been involved in the project "for about ten years, on and off," and he continues to lead the development of design criteria and reviews all construction engineering drawings.

The bridge's newly constructed towers now rise 510 feet above mean sea level, with each tower consisting of 8,000 cubic yards of concrete. Some people have expressed a wish to see the concrete towers painted in the same "Narrows Green" as the 1950 bridge. This, however, would be an expensive proposition, requiring repainting every five or ten years. Such a major maintenance expense does not have much support. The current and less expensive plan is to apply one coat of "Washington Gray," a pigmented sealer that would be allowed to age naturally.

The new bridge is 5,400-feet long, with a 2,800-foot main span. The deck will carry four 11-foot-wide lanes for eastbound traffic toward Tacoma. Two lanes are for general purpose traffic, one is a high occupancy vehicle (HOV) lane, and the fourth is an "add/drop" lane that will run between a new toll plaza on the west side of the bridge and Jackson Avenue on the east side of the span. There is a two-foot inside shoulder, 10-foot outside shoulder, and a 10-foot bicycle/pedestrian path separated by a barrier. The foundations, towers, stiffening truss, and anchorages are designed for adding a second deck (either roadway or light rail) to the bottom of the truss in the future.

The 1950 bridge will carry traffic westbound to the Peninsula. One of the changes ahead is a major "retrofit" of the 1950 structure. To improve safety, three of the five lines of wind grating in the roadway will be removed. This, plus renovating the roadway, will create three traffic lanes (two general purpose and one HOV), increasing each lane from 9½ to 12 feet, the current standard lane width. Work on the 1950 bridge also will include seismic protection. In addition, the project includes improvements to 3.4 miles of SR 16 from the Jackson Avenue interchange in Tacoma to west of a new 36th Street NW half-diamond interchange.

Once the 2007 bridge is finished, it and the 1950 span will belong to an exclusive "twin suspension bridges club." There are only four other sets like them in the world—twin bridges across the Bosporus connecting Europe and Asia, the Delaware and New Jersey Memorial twins, Minnesota's Brumley pair, and the William Preston Lane Jr. bridges in Washington, D.C.

COSTS AND BENEFITS

About $800 million of tax-exempt bond financing has been required for the project, while the State has funded $50 million and provided certain tax exemptions and deferrals that help to reduce the project costs. The new bridge construction and the 1950 span's improvements will cost a total of $615 million, which will be repaid by toll revenue. The project is a fixed-price project, thus costs must remain within that $615 million amount. The total project budget is $849 million, including improvements to State Route 16 and the seismic retrofitting of the 1950 bridge. Upon opening the new bridge in 2007, WSDOT will begin collecting tolls from eastbound motorists. The exact toll amount will be determined by the Washington Transportation Commission after receiving recommendations from a nine-member Citizen Advisory Committee appointed by the Governor. Citizen committee members must reside within the project area.

In regard to benefits to drivers, safety is foremost. Dividing traffic onto separate westbound and eastbound bridges will eliminate opposing traffic lanes, thus significantly improving safety. Additional safety improvements include wider lanes and new safety shoulders.

A second benefit will be less traffic congestion. The operational improvements along the SR 16 corridor between the Nalley Valley viaduct in Tacoma and the Olympic Drive interchange in Gig Harbor will help reduce delays. Although the new bridge is not intended to eliminate all gridlock in the area, it will relieve the severe traffic congestion experienced at the Narrows today.

OF TIME AND BRIDGES

Tides and time flow swiftly through the Narrows, linking generations. The 1940 and 1950 bridges still send ripples, even today contributing new knowledge in engineering research. Once again, with a new span under construction, ironworkers dare danger by clambering on high steel and concrete.

"That new tower is quite a sight," commented retired ironworker Earl "Whitey" White as he stood atop the 1950 bridge's east tower on a sunny afternoon in August 2005. White had just finished walking up the north cable, not something many octogenarians would attempt. He looked down thoughtfully at cars whizzing over the span: "This is really a highlight of my later years."[2]

More than a half-century ago, when the 1950 bridge was under construction, White traveled over that cable hundreds of times. He often worked graveyard shift, arriving and parking his car at the east anchorage before midnight. After punching the time clock and reporting for work, he had to get to the west anchorage, and the only way was to take a mile-long hike on the cable. After working through the dark and completing his shift at daybreak, he walked over the cable again to locate his car and drive home. It was just another day on the job.

On Whitey's latest journey up the north cable, he set a leisurely pace. He stopped a few times to catch his breath. But mostly he wanted to enjoy the experience; to drink from a glass of his past with gusto and awareness. Part of that enjoyment came from chatting with his companions, WSDOT's Kip Wylie and Sharan Linzy, and Dan Weber, a structural steel inspector for a private firm. They talked about the balls of herring that swim with the Narrows tides, and how they are so much smaller now. White reminisced about his World War II experiences in the Marine Corps, and marveled at the progress on the new 2007 bridge; its east tower stood only yards away. Finally, the cable strollers reached the tower top, climbed around the cable saddle, and stepped onto the walkway between the tower legs.

This small but historic moment eventually ended as quietly as it began, but its significance was not lost on White or his companions. Today, there are a mere handful of men alive who worked on the 1950 bridge. Earl White, Arnie Colby, and Bill Matheny are among the exclusive group. They may not have fully realized it at the time, but they cre-ated more than a careful arrangement of steel and concrete. They built an extraordinary machine, one that would continue for many years to connect people over a vast area.

Today, there is great synergy in the twin bridges. They are unique structures, honored mainly for their practicality and efficiency, yet their magnificent bold profiles on the south Puget Sound horizon inspire and uplift us. They are, as was Galloping Gertie, the embodiment of a dream.

The Tacoma Narrows bridges, children of years of trial, error, tragedy, and persever-ance, are brilliant triumphs of the human mind and spirit.

Notes

1. Tim Moore, interview, September 2005; "Tacoma Narrows Bridge Models Tested in NRC Wind Tunnel," press release, May 2, 2003, National Research Council (Canada), web site, www.nrc-cnrc.gc.ca/highlights/2003/0306tacoma_e.html.
2. Earl White, interview, August 23, 2005.

Appendices

APPENDICES

Statistical Comparison of the 1940 and 1950 Tacoma Narrows Bridges

General	1940 Bridge	1950 Bridge
Cost	$6,618,138	$14,011,384
Total structure length	5,939 feet	5,979 feet
Suspension bridge section	5,000 feet	5,000 feet
Center span	2,800 feet	2,800 feet
Shore suspension spans (2), each	1,100 feet	1,100 feet

Suspended Structure

	1940 Bridge	1950 Bridge
Roadway height above water (mean sea level)	190 feet at towers, 208 feet at center span	200 feet at towers, 212 feet at center span
Center span vertical clearance above water	203 feet	184½ feet
Weight of center span	5,700 lb./ft.	7,250 lb./ft
Traffic lanes	2	4
Width between cables	39 feet	60 feet
Width of sidewalks (2), each	5 feet	3 feet 10 inches
Width of roadway	26 feet	49 feet 10 inches
Thickness of roadway	5¼ inches, reinforced concrete	6⅜ inches, reinforced concrete
Suspender cables, intervals	50 feet	32 feet
Number of girders and type	2 plate girders	2 Warren trusses
Depth of stiffening girder	8 feet	33 feet
Ratio, deck width to center span	1:72	1:46
Ratio, deck depth to center span	1:350	1:85

Anchorages

	1940 Bridge	1950 Bridge
Weight of each anchorage	52,500 tons	66,000 tons
Concrete in each anchorage	20,000 cu. yds.	25,000 cu. yds.
Structural steel in both anchorages	592 tons	901 tons
East approach and anchorage	345 feet	365 feet
West approach and anchorage	594 feet	614 feet
West anchorage (concrete anchor block and gallery)	164 feet long	169 feet long

East anchorage (concrete anchor block and gallery), approach, administration buildings, and toll house	173 feet long	185 feet long
West anchorage, construction and cost	Woodworth & Cornell $299,734	Woodworth & Co. $406,000 (est.)
East anchorage, construction and cost	Pacific Bridge Co. $272,273	Woodworth & Co. $386,000 (est.)
Cable anchor bars	38 in each anchorage 54 feet long	38 in each anchorage 62 feet long

Cables

Diameter of main suspension cable	17½ inches	20¼ inches
Weight of main suspension cable	3,817 tons	5,441 tons
Weight sustained by cables	11,250 tons	18,160 tons
Sag ratio	1:12	1:10
Number of wire strands in each cable	19	19
Number of No. 6 wires in each strand	332	458
Number of No. 6 wires in each cable	6,308	8,705
Total length of wire	14,191 miles	19,715 miles
Cost of cables, bands, suspenders, fittings, etc.	$1,335,960	$2,732,773

Towers

Height above water (mean sea level)	448 feet	507 feet
Height above piers	425 feet	467 feet
Height above roadway	230 feet	307 feet
Weight of each tower	1,927 tons	2,675 tons
Cost of both towers	$542,144	$1,977,167

Piers

East Pier (#5), total height	247 feet	265 feet
East Pier (#5), depth of water	135 feet	135 feet
East Pier (#5), penetration at bottom	90 feet	90 feet
West Pier (#4), total height	198 feet	215 feet
West Pier (#4), depth of water	120 feet	120 feet
West Pier (#4), penetration at bottom	55 feet	55 feet
Area	118 feet 11 inches by 65 feet 11 inches	118 feet 11 inches by 65 feet 11 inches

SPANNING TIME: A CHRONOLOGY OF THE TACOMA NARROWS BRIDGES

1792

- On May 20, Archibald Menzies, botanist with the Captain George Vancouver expedition, notes "a most Rapid Tide" when passing through the Narrows.

1888 or 1889

- John G. Shindler, a rancher traveling through the Narrows on Ed Lorenz's steamboat, declares, "Captain, some day you will see a bridge over these Narrows." Lorenz later says, "We all thought Shindler was crazy" (*Tacoma Times*, September 12, 1939).

1889

- A railroad crossing (probably a trestle, rather than a bridge) at the Narrows is briefly contemplated by the Northern Pacific Land Department. A clerk named George Eaton proposes the link between the terminus of the transcontinental line in Tacoma and Port Orchard, a proposed site for the Puget Sound Naval Shipyard.

1923

- The Federated Improvement Clubs of Tacoma launch a campaign for a bridge between Point Defiance and the Gig Harbor area, requesting support from civic groups in surrounding communities. In late December, C.F. Mason, a realtor and president of the organization, says they have been working on the Narrows proposal "for some months."

1926

- Llewellyn Evans, a Tacoma City Utilities Department superintendent and the local Good Roads Association president, receives endorsement from the Tacoma Chamber of Commerce to begin a campaign for a Narrows bridge.
- Pierce County grants a 10-year contract (including a guarantee of immunity from competition) to Mitchell Skansie for a ferry service across the Narrows. Skansie organizes the Washington Navigation Company.

1927

- The Tacoma Chamber of Commerce appoints John Baker as chairman of a committee to guide efforts to build a Narrows bridge.
- J.F. Hickey, president of the Tacoma Chamber of Commerce, tells the *Tacoma Times*, "Within a few weeks a survey will be underway and…shortly after January we will know whether the immediate and future business of a large portion of the Olympic territory will, hereafter, drain into Tacoma, as a result of this important connecting link."
- Skansie's Washington Navigation Company begins ferry service.
- **June.** The Roads Committee of the Tacoma Chamber of Commerce estimates that a Narrows bridge will cost between $3 and $10 million. Tacoma newspapers provide helpful publicity and editorial support.
- **September.** Joseph B. Strauss, a noted bridge engineer from Chicago (and later

builder of the Golden Gate Bridge), makes a visit regarding the proposed Narrows bridge.

- **November.** H.H. Meyers of the New York-San Francisco Development Company visits the area to discuss the proposed project. Meyers estimates the cost at between $7 and $15 million.

1928

- **August.** Charles A. Cook, a local realtor and civic activist, proposes that Pierce County build a steel cantilever bridge. Cook suggests a structure similar to the Carquinez Strait Bridge, a 4,500-foot span costing $8 million then under construction about 20 miles north of Berkeley, California.
- **November.** The Tacoma Chamber of Commerce hires noted bridge architect David B. Steinman of the New York firm Robinson and Steinman to conduct preliminary work for a proposed bridge. Steinman visits Tacoma and, over the next two years, will spend $5,000 on a preliminary survey, traffic estimate, reports, and layout, design, and architectural drawings.
- Tacoma's 6th Avenue Commercial Club rallies public support.
- **November.** Survey work begins and soundings are taken in the Narrows.
- **December.** The Tacoma Chamber of Commerce supports proposed legislation to provide a state franchise for building a bridge. City of Tacoma leaders and the Pierce County Board of Commissioners formally ask Washington State officials to construct a Narrows bridge.

1929

- **February.** The Washington State Legislature passes a law authorizing a Narrows bridge. Three Tacoma Chamber of Commerce members will be granted a franchise for a toll facility.

- **March.** Architect David B. Steinman makes a second visit to Tacoma. A proposed bridge design by Steinman, imposed on a Narrows photo taken by M.D. Boland, appears in local newspapers on March 5. Steinman's proposed suspension span would measure 4,944 feet in length with towers rising 670 feet above the water. The design features a 2,400-foot center span, two side spans of 912 feet, and another approach span of 720 feet on the west side connecting to the Peninsula. Steinman estimates the cost at $9 million.

1931

- The Tacoma Chamber of Commerce decides that David Steinman's firm is "not sufficiently active" in obtaining financing and asks Steinman to cancel their agreement. Steinman decides the estimated revenues from traffic tolls would be too low to justify construction.
- Tacoma city engineers propose a steel cantilever truss bridge carrying railroad traffic as well as motor vehicles and pedestrians, costing a shocking $12 million. The plan calls for five spans on four piers, with a 54-foot-wide roadway allowing for two automobile lanes and a railroad track in the center.

1932

- The Tacoma Chamber of Commerce signs a contract with Elbert M. Chandler of Olympia to build a proposed "suspension" bridge having a 1,200-foot central span, a vertical clearance of at least 196 feet, a 24-foot-wide deck for two highway lanes, and costing not more than $3 million. Chandler requests a loan from the Reconstruction Finance Corporation (RFC) to be repaid by user tolls. The federal RFC refuses funding to buy Mitchell Skansie's ferry system.

- Leon F. Moisseiff, consulting bridge engineer with Moisseiff & Associates of New York, reviews Chandler's proposal and reports that the bridge is feasible.
- **November.** Chandler's revised plan now proposes a 7,000-foot-long steel "cantilever" with a 1,200-foot central span and six spans of about 600 feet each (plus approaches). The bridge would have 10 piers, 2 on land, 2 out of water at low tide, and 6 in about 150 feet of water on either side of the steel cantilever.

1933

- **November.** The U.S. Navy and U.S. Army declare support for Chandler's proposal, advocating federal approval because of the military necessity for an improved route between the Bremerton Naval Shipyard and Fort Lewis.

1934

- Two Tacoma politicians, U.S. Senator Homer T. Bone and U.S. Representative Wesley L. Lloyd, introduce legislation in Congress to provide federal funds for building a Narrows suspension bridge.
- Encouraged by the creation of the Public Works Agency (PWA), providing federal funds under President Roosevelt's "New Deal" program for major public construction projects, Pierce County government leaders have renewed hope. They apply for a grant for Chandler's plan in 1933, but the federal bureaucracy declines funding in October 1934.

1936

- **January.** The "Narrows Bridge Gang," a coalition of Tacoma community groups and businessmen headed by Wallace Morrisette, begins a statewide letter writing campaign to persuade President Roosevelt to support funding for a Narrows bridge.

- **January 22.** The War Department advocates a revised application by Pierce County Commissioners. The new plan calls for a suspension span (instead of a cantilever type). Pierce County officials prepare an application for $4 million, with 45 percent coming from a PWA grant and 55 percent paid by Pierce County public utility bonds. Included are funds to purchase the existing ferry system.
- **March.** After three months of negotiations, Elbert Chandler and other private firms with an investment in Narrows bridge planning sell their interest to Moran & Proctor, a New York engineering firm.
- **March 29.** Tacoma newspapers report that Moran & Proctor have prepared preliminary plans for a suspension bridge costing $4,089,091. The 4,944-foot-long structure would feature a center span of 2,400 feet and two side spans, 912-feet each.

1937

- **January.** The Washington State Legislature creates the Washington Toll Bridge Authority (WTBA), patterned on a California law. The legislature appropriates $25,000 to study the Tacoma-Pierce County request to build a bridge. Pierce County transfers its application for constructing a bridge to the WTBA.
- **October 1.** Tacoma boosters present a pamphlet proposing a Narrows bridge to President Franklin D. Roosevelt, who is visiting the city.

1938

- **May 5.** Pierce County deeds Tacoma Field to the U.S. government, thus establishing McChord Field. In late summer, a $5 million construction project begins to improve the air base, employing some 2,000 men.

- **May 23.** The WTBA submits an amended application to the PWA and applies for a RFC loan. The revised application includes a preliminary layout design by state engineer Clark Eldridge for a suspension bridge.
- **June 23.** The Public Works Administration grants funding, culminating more than 14 years of community efforts. The PWA award is conditional on the WTBA hiring outside design consultants—Leon Moisseiff for the superstructure, and Moran & Proctor of New York for the substructure.
- **June 27.** The WTBA accepts the PWA grant offer of $2.7 million and an RFC loan of $3.3 million.
- **July 18 and July 27.** Moisseiff submits reports to Director Lacey V. Murrow of the Washington State Highway Department with recommendations for changing Clark Eldridge's superstructure plan.
- **August.** Moisseiff completes the revised drawings.
- **September 27.** Construction bids are opened, with the Pacific Bridge Company taking the contract with a low bid of $5,594,730.40. Associate contractors include Bethlehem Steel for supplying steel and John A. Roebling's Sons of New York for the wire.
- **October.** After the WTBA has received bids from construction firms, all of which were above the budgeted projections, the State re-applies to federal authorities for more funds. The PWA increases its grant to $2,880,000 and the RFC to $3,520,000, for a total of $6.4 million to build the bridge.
- **November 23.** Beginning of construction, though the official contract "start date" is two days later.

1939

- **January 25.** First 520½ ton anchor for west pier construction is dumped into the Narrows.
- **February 24.** Caisson anchors are placed for the west pier (#4).
- **March 18.** Caisson for the west pier (#4) arrives.
- **March 19.** First anchor is secured to the west caisson.
- **April 18.** Excavation begins for the east anchorage.
- **May 2.** Excavation begins at the west anchorage.
- **May 9.** Caisson for the west pier (#4) reaches the bottom of the Narrows.
- **May 12.** Excavation completed for the east anchorage.
- **May 15.** Caisson for the east pier (#5) arrives. Excavation is completed at the east anchorage.
- **May 18.** Excavation completed for the west anchorage.
- **May 19.** West pier (#4) secured to all anchors and grounded on the bottom of the Narrows.
- **June 9.** East pier (#5) is secured to all anchors and grounded on the Narrows bottom.
- **July 13.** West pier (#4) is completed and ready for tower steel.
- **July 15.** Cable anchor bars are completed at the east anchorage.
- **July 22.** Cable anchor bars are installed at the west anchorage; work on both anchorages is suspended pending completion of the cable spinning.
- **August 2.** Bethlehem Steel begins erection of the west tower (#4).
- **September 15.** Completion of the east pier (#5).
- **October.** Catwalks and structures for cable spinning are started.
- **November.** Bethlehem Steel completes the east tower (#5).
- **November 6.** Parker Painting Company, a subcontractor, begins sand blasting and painting the outside of the west tower (#5).
- **November 13.** At 7:45 A.M., the strongest earthquake in decades strikes the region.

The 6.2 magnitude jolt, with an epicenter some six miles south of Bremerton, rumbles and shakes the Narrows Bridge. Engineers report no damage to the structure.

1940

- **January 10.** Cable spinning begins.
- **February.** The WTBA hires Professor F. Burt Farquharson, who in the ensuing months, with aid from students, will build models of the bridge deck for $20,000. Wind tunnel tests on the models at the University of Washington's Engineering Experiment Station begin when Galloping Gertie's "bounce" becomes quite apparent by mid 1940.
- **March 5.** Cable spinning completed.
- **March 21.** First section of steel deck, 111-feet long and weighing 91 tons, is hoisted into place.
- **April 12.** First pour of approach roadway concrete.
- **April 18.** Painting of the main cables, suspenders, and fittings begins by subcontractor Fisher & White Company of Seattle.
- **May 6.** Completion of steel floor system (girders, beams, and stringers); finishing rate was 200 feet per day. About this time, riveters and other workmen notice the "bounce," or "galloping," of the bridge. Some chew on lemons to combat nausea.
- **May 17.** Concrete pouring of roadway begins.
- **May 21.** Completion of concrete pouring for the center of the roadway, at 300 feet per day.
- **May (late).** Four hydraulic jacks are installed to act as shock absorbers as engineers hope to take the "bounce" out of the bridge, but it has no effect.
- **June 10–21.** Catwalks dismantled.
- **June 27.** Only three days before the official completion and public opening, the first and only death during construction occurs when carpenter Fred Wilde stumbles and falls 12 feet.

- **June 28.** A lucky bridge worker falls 190 feet into the Narrows and survives. Pete Kreller, a 26-year old painter, sustains relatively minor injuries. Bridge engineers tell Kreller that his fall lasted 4 seconds and he reached a speed of 60 mph.
- **June 30.** Completion of the concrete roadway, sidewalks, and curbs on the suspended structure.
- **July 1.** Official opening ceremonies. Construction time was 19 months. Engineers announce that there is nothing dangerous in the bridge's "bounce."
- **July 23.** The first airplane pilot, defying the obvious danger, flies under the bridge.
- **July–September.** The bridge's roadway sometimes "bounces" in a wind as light as 4 mph. Waves in the deck ranging from 1 to 5 feet (a total rise and fall of 2 to 10 feet) are common. In a couple of extreme cases, 10-foot waves (a total rise and fall of 20 feet) are experienced, making some motorists "seasick."
- **September 11.** A falling bucket during the final paint work kills bridge painter Hugh Meiklejohn.
- **October 1.** Completion of the application of three coats of green paint. Total paint required, 5,800 gallons; total painting payroll, 37,200 hours; and total steel painted, 13,000 tons.
- **October 4 and 7.** Bridge engineers add temporary "tie-down" cables to the side spans to try to reduce Gertie's "bounce." Farquharson and State bridge engineers believe they have the problem solved.
- **November 1.** A tie-down cable on the east side span breaks in a high wind when Gertie begins to "gallop." Workmen immediately replace the cable.
- **November 2.** Farquharson completes wind tunnel studies on a 54-foot-long scale model. He discovers a twisting motion that could potentially destroy the span. Farquharson informs the WTBA that the probable cause of Gertie's "ripple" is the solid stiffening girders, which catch

the wind and make the bridge susceptible to aerodynamic forces. State authorities begin drafting a contract to have wind deflectors installed.

- **November 7.** "Galloping Gertie" begins collapsing at 11:02 A.M. when a 600-foot section of roadway in the western half of the center span (the "Gig Harbor quarter point") breaks free. At 11:08 A.M., a final section of the center span falls into the Narrows.
- **November.** Planning for dismantling and salvage operations begin.
- **December 2.** Insurance agent Hallet R. French of Seattle is arrested for grand larceny after pocketing a $70,000 premium for an $800,000 policy written with the State for the Narrows Bridge.
- **December.** Contract for start of the removal of the remaining 1940 bridge superstructure is awarded to J.H. Pomeroy & Company. The "Carmody Board," appointed by the Federal Works Agency, begins an investigation of the bridge failure.

1941

- **February 7.** Hallet R. French is sentenced to 15 years in the Washington State Penitentiary at Walla Walla. He will serve two years, then be released for "good behavior." Hallet's next job is at a Seattle shipyard
- **March 1.** Washington State files insurance claim for $5,200,000.
- **March 7.** Isaac F. Stern of Chicago is appointed to a three-man arbitration board to represent the insurance underwriters.
- **March 28.** Final report issued by the "Carmody Board."
- **April 20.** Bridge engineer Clark Eldridge resigns from the State Highway Department and takes a position with the U.S. Navy on Guam.
- **June 2.** The insurance underwriters file their report. The piers, cables, and tow-

ers all could be salvaged and reused, they claim, and offer the State a settlement of $1.8 million.
- **June 26.** A State investigating board files its report stating that the bridge is virtually a total loss, except for the piers; the total insurance claim is $4,297,098.
- **July 15.** Dexter R. Smith begins work as design engineer for a new bridge (1950) under the WTBA's Charles E. Andrew.
- **August 26.** The State and 22 insurance companies agree on a $4 million settlement for the near-total loss of the 1940 bridge.
- **September.** Start of dismantling of the bridge structure.

1942

- **August.** Contract for removal of the cables and towers is awarded to Philip Murphy & Woodworth Company.
- **November 7.** Salvage of steel cables begins on the second anniversary of Galloping Gertie's collapse.
- **December 1.** Insurance embezzler Hallet French, recently released from the State Penitentiary, is working at a Seattle shipyard.

1943

- **June.** Dismantling of the first Narrows Bridge is completed. The WTBA paid $646,661 for the salvage operation, which brought a meager return of $295,726 for 7,000 tons of scrap steel.
- **September 3.** Leon Moisseiff dies of heart failure at his summer home in Belmar, New Jersey, at age 71.

1944

- **June 15.** The Washington Toll Bridge Authority adopts a design for a new Narrows Bridge.

1946

- **March 19.** Governor Mon G. Wallgren announces that revised plans for the new Narrows Bridge have been finalized and that negotiations for insurance on the span are opened.
- **April.** Revised designs approved. The projected cost is $8.5 million for 22 months of construction work.

1947

- **January.** An independent consulting firm, Modjeski & Masters of Harrisburg, Pennsylvania, confirms that the 1940 piers will meet all requirements for the proposed new span.
- **April 30.** Governor Wallgren announces that insurance for the bridge has been arranged, with 100 companies participating. Models aerodynamically verify the soundness of the design.
- **August.** The State requests bids; the cost has risen to $11.2 million.
- **October 15.** Bids are opened and contracts are awarded to Bethlehem Pacific Coast Steel Corporation ($8,263,904.13 for the superstructure) and John A. Roebling's Sons Company of San Francisco ($2,932,681.27 for cable work). However, financing is not yet arranged and the start of construction is delayed.
- **December.** The WTBA offers a bond issue of $14 million to finance construction.

1948

- **March 12.** Bond financing is completed.
- **March 31 and April 1.** Contracts are granted to primary contractors Bethlehem Pacific Coast Steel and John A. Roebling's Sons to construct the replacement Tacoma Narrows Bridge.

- **April 9.** Construction begins with site clearing and earth moving at the east anchorage.
- **May 24.** Robert E. Drake, a carpenter at the west anchorage, becomes the first worker to die building the 1950 bridge.
- **December 15.** Work completed on the east pier (#5).
- **December.** The west pier (#4) is finished and tower erection starts with placement of base plates by Bethlehem Steel.

1949

- **April 12.** Construction of the east tower (#5) is nearly complete, except for cable saddles.
- **April 13.** An earthquake measuring 7.1 on the Richter scale shakes the Puget Sound region. The piers and the towers (then under construction) suffer no damage. During the quake, the towers sway as much as six feet from perpendicular. The 28-ton cable saddle on the north side of the east tower (#5) was in place, but not secured. It fell 500 feet, plunging through and sinking a barge, and settled 140 feet below the Sound. The cable saddle is retrieved in three days, and after another week is repaired and returned to its perch atop the tower.
- **June 1.** East tower (#5) is 98 percent complete, with more than 71,000 rivets driven. A Chicago boom is hung at roadway level on the south tower leg in preparation for raising truss steel.
- **June 8.** Fire destroys the creosote timber fender of the west pier (#4), but the west tower suffers only minor damage. Heat from the flames damage some cable spinning machinery, 400 feet above. The fender is later rebuilt.
- **June 14.** Riveting at the east tower is completed, with a total of 71,700 rivets driven.
- **June (late).** Effective completion of both east and west towers.

- **June 30.** Foss Tug No. 11 pulls the first line from the east anchorage to the east tower. The line is used to install pullback and catwalk cables.
- **July 17.** Official completion date for both the east and west towers. Work is turned over to John A. Roebling's Sons Company for cable construction.
- **September 15.** With completion of the catwalks, preparations begin for cable spinning. First man to walk the entire distance over the Narrows on the catwalk is Harold Hills, a field engineer for Roebling's Sons.
- **September 16.** Harry Cornelius, inspector for the WTBA, became the second person to stride over the entire catwalk. His time was 32 minutes, apparently faster than that posted the day before by Harold Hills.
- **October 13.** Guide wire for spinning operations is set.
- **October 17.** Start of cable spinning, with the first wire of the first strand for the north cable. Roebling's crew numbers 184 men at the bridge site and 7 men at the Tacoma reeling plant.
- **October 26.** Completion of the first strand for the north cable. Work begins on strand no. 1 of the south cable. Roebling's crew now includes more than 220 men, with 60 men at the reeling plant, all working two shifts.
- **November.** Roebling's runs three shifts for cable spinning, employing more than an average 250 men at the bridge, with 70 at the reeling plant working two shifts. Work sometimes slows due to high winds.

1950

- **January.** Roebling's Sons drop the first suspender cables for connecting the main cables to the deck.
- **January 16.** Completion of cable spinning for main suspension cables.
- **March 7.** Cable band placement for both cables is completed.

- **April 7.** At 3:45 P.M., 36-year-old Lawrence Stuart Gale, an ironworker, dies when a weld gives way during deck construction and he plummets 180 feet into the Narrows. He is the second worker to be killed on the bridge project.
- **May 21.** Ironworkers place the closing top chord of steel on the bridge deck at mid-span.
- **May 28.** Ironworkers finish closing the deck on the east side.
- **May 29.** Ironworkers close the deck on the west side.
- **May 31–June 1 (midnight).** Final closing of the deck between east and west.
- **June 6.** Just after the start of the morning shift, Glen "Whitey" Davis dies when falling 190 feet into the Sound.
- **June 25.** Start of concrete pouring on the road deck.
- **July 24.** Completion of cable wrapping and caulking of cable bands; handrails next to sidewalks installed.
- **July 31.** Ray Bradley, a welder employed by Bethlehem Pacific Coast Steel, dies from a heart attack or electrocution.
- **August 22.** Concrete pours for the roadway slab and painting of steel parts are in full swing. Some 110 men are employed on the concrete pours, setting forms and laying reinforcing steel in the roadway. Seven painters are working on the main cables, suspender cables, and cable bands. Eleven men are sandblasting and painting the truss.
- **September 30.** Completion of the installation of roadway grating and mid-span hydraulic jacks. Roebling's Sons has more than 50 men spot painting on the cables and dismantling working platforms, catwalks, and derricks. More than 100 men are working on various concrete pours. Sidewalks and road deck are nearing 100 percent completion. Final work on the toll plaza facilities, such as plastering and plumbing, is underway.

- **October 14.** After 30 months of construction work, a new and much safer Tacoma Narrows Bridge opens. Tolls are 50¢ for a car and driver, one-way, 10¢ per passenger.

1951

- **January–March.** Several storms with sustained winds up to 75 mph sweep through the Narrows. The bridge stands solid, showing no vertical or torsional movement and only a slight lateral deflection.
- **November.** Completion of all components of the 1950 Narrows Bridge.

1952

- **October.** Establishment of the Living War Memorial Park, located on the south side of the east approach.
- **November 11 (Armistice Day).** Dedication of the Living War Memorial Park, honoring Tacoma and Pierce County soldiers who served in World War II.

1954

- All bolts in the suspension cable bands are overhauled. Four workmen remove, clean, and replace 2,200 bolts, one at a time, in a yearlong effort.

1965

- **May 14.** Tolls are officially removed in a 10:00 A.M. ceremony.

1983

- **June 23.** A 19-year-old woman becomes the first to survive a fall from the bridge.

1993

- The State Legislature unanimously enacts HB 1006, the Public Private Initiatives (PPI) Act (RCW 47.46), as a way to secure new financing for improving the state's transportation infrastructure.

1994

- The legislature unanimously enacts SHB 2909, authorizing $25 million of general obligation bonds to provide state financial participation for PPI programs and projects.

1995

- **January.** WSDOT issues a competitive solicitation to private companies worldwide. Three different private sector teams propose improvements to the Tacoma Narrows Bridge and the State Route 16 (SR 16) corridor. United Infrastructure Company is selected and approved by the State Transportation Commission to negotiate an agreement to finance, develop, construct, and manage the Tacoma Narrows Bridge Project.
- The State Legislature amends RCW 47.46 with passage of 3ESHB 1317. This bill requires that prior to executing a PPI agreement, WSDOT must conduct a public advisory election if there is opposition to a project, evidenced by the submission of petitions bearing 5,000 signatures. The Peninsula Neighborhood Association submits the requisite signatures, meeting the advisory election requirements. The law also proscribes a public involvement process, and specific criteria for establishing a regional boundary for the vote.

1996

- Senator Oke introduces SB 6509 to provide $500 million of general obligation bonds for the Tacoma Narrows Bridge effort, which does not pass out of the Senate Transportation committee. Instead, Senator Oke's ESSB SB 6753 is enacted, requiring that before WSDOT conducts an advisory election on a PPI project, the legislature must authorize funding for environmental and engineering studies, public involvement activities, and to pay for a regional advisory election. The legislature appropriates approximately $11 million for the Tacoma Narrows Bridge Project.

1997

- The Major Investment Study identifies 22 potential multi-modal alternatives for efficient SR 16 transportation.

1998

- The legislature enacts SHB 3015 to provide sales tax deferrals and limited tax exemptions for the Tacoma Narrows Bridge Project. At the time, this would in effect provide over $74 million in tax relief to toll payers. Following the Major Investment Study, the Draft Environmental Impact Statement, and numerous public involvement activities, WSDOT recommends the preferred alternative, which includes the addition of an HOV lane in each direction from the Kitsap/Pierce county line connecting to the I-5 HOV lane system, approximately 14 miles. The recommendation also calls for the construction of a new parallel bridge south of the 1950 bridge, reconfiguration of the existing bridge, and other roadway improvements. The advisory election is held in all or parts of a seven county area with 53 percent voting in favor of the project.

1999

- The State Legislature authorizes a $50 million contribution to the Tacoma Narrows Bridge Project. The Secretary of Transportation and the United Infrastructure Company execute an "Agreement to Develop, Finance, Construct and Operate the Tacoma Narrows Bridge Project" on June 15, 1999.

2000

- The legislature appropriates $50 million, following passage of I-695. The State Supreme Court rules that the 1960s statute prohibits collection of tolls for improvements to the existing bridge, unless there is authorizing legislation.
- A thorough inspection by a private firm marks a milestone for the 1950 bridge. The final report concludes that the "bridge is one of the best in the nation for its maintenance and condition."

2001

- The legislature fails to resolve the conflicting state laws. Instead, it debates over a state-financing plan, involving the issuance of general obligation bonds for covering the project's debt. After seven months of deliberations, the legislature adjourns in July without action. Consequently, the project stands down and is almost completely demobilized, resulting in millions of dollars of delay costs.

2002

- The State Legislature enacts legislation calling for over $800 million in state bonds to be issued and repaid with tolls. The business agreement with the United Infrastructure Constructors is terminated and they are compensated for their costs to date. WSDOT assumes management of the project operation and construction.
- **July 16.** WSDOT executes a $615 million fixed-price-schedule, design-build agreement with Tacoma Narrows Constructors (TNC), a joint venture of Bechtel Infrastructure and Peter Kiewit Sons. Completion of the new Narrows toll bridge is expected in about 5½ years.

- **September 18.** First bond sales for the initial project funding.
- **September 25.** WSDOT issues notices to proceed to TNC and TransCore.
- **October 5.** The official start and public ground breaking ceremony for the new Tacoma Narrows Bridge construction, with completion scheduled for 2007.

GLOSSARY

Glossary

Abutment—Substructure element supporting each end of a single or multi-span superstructure, and, in general, retaining or supporting an approach embankment.

Anchor span—A span that counterbalances and holds in equilibrium the cantilevered portion of an adjacent span during construction.

Anchorage—Massive concrete structure, also called a "cable anchorage," providing stability where the cable ends are tied and withstanding the tremendous stress of the loaded cables.

Approach span—A span or spans connecting an abutment with the main span or spans.

Beam—Linear structural member designed to connect from one support to another.

Bent—A substructure unit supporting two ends of a bridge span; also called a pier. Made up of two or more columns or column-like members, connected at the top by a cap, strut, or other member.

Box girder—Support beam having a hollow box form, which in cross-section is a rectangle, square, or trapezoid.

Cable—The part of a suspension bridge or cable-stayed bridge providing support to the suspended spans. Consists of many steel wires bound together into strands and anchored at each end.

Cable saddle—A cable saddle sits at the summit of a tower, channeling a main suspension cable where it crosses over the top of a tower leg. As temperature changes, wind, or traffic affect the movement of cables, saddles absorb the load, shifting it to the towers.

Cable spinning—The technique of pulling wires with a "spinning wheel" or "traveler" from an anchorage over a tower and back again to form a main cable. The wires are grouped into strands, then bound tightly together to form strong suspension cables. At the anchorage, the strands are attached to eye-bars.

Cable-stayed bridge—A bridge in which the superstructure is directly supported by cables or stays, passing over or attached to a tower or towers located at a main pier or piers.

Caisson—(French term for "box") A caisson is a huge box typically consisting of steel-reinforced, waterproof concrete with an open central core. The base of a caisson has a "cutting edge" of plate steel for penetration. In a suspension structure, a caisson becomes the foundation (i.e., the pier), supporting bridge towers.

Cast-in-place—Concrete poured into an on-site formwork to create a structural element in its final position.

Catwalks—Temporary foot bridges used by bridgeworkers to spin main cables (several feet over each catwalk) and to attach suspender cables connecting the main cables to the stiffening truss.

Chord—A primary horizontal member of a truss extending along the length of a span.

Cross brace—Transverse brace between two main longitudinal members.

Damping—The action of reducing the vibration of an object. This tends to return the vibrating object to its original position.

Dead load—Static load due to the weight of the structure itself.

Deck—Roadway portion of a bridge directly supporting vehicular and pedestrian traffic.

Deck bridge—A bridge in which the supporting members are all beneath the roadway.

Deck truss—A bridge with a roadway supported from beneath by a truss.

Diagonal—Sloping structural member of a truss or bracing system.

Eye-bar—Steel bars holding the wire strands of the main cable. Eye-bars are attached to beams embedded in the concrete of an anchorage.

Expansion joint—A joint designed to allow for expansion and contraction movements produced by temperature changes, load, or other forces.

Fatigue—Cause of structural deficiencies, usually due to repetitive loading over a long period of time.

Flutter—Self-induced harmonic motion, producing aerodynamic instability that can grow to large amplitudes of vibrations.

Footing—The enlarged, lower portion of a substructure that distributes the structure load either to the earth or to supporting piles. The most common footing is a concrete slab. "Footer" is a colloquial term for footing.

Girder—A main support member usually receiving loads from floor beams and stringers; also, any large beam, especially if built up.

Hanger—Tension member serving to suspend an attached member.

Hinge—A point in a structure at which a member is free to rotate.

Joint—In stone masonry, the space between individual stones; in concrete, a division in the continuity of the concrete; in a truss, the point at which members of a truss frame are joined.

Live load—Vehicular traffic.

Lower chord—Bottom horizontal member of a truss.

Main beam—A beam supporting the spans and bearing directly onto a column or wall.

Member—An individual angle, beam, plate, or built piece intended to become an integral part of an assembled frame or structure.

Moment—Force exerted on a component of a bridge.

Oscillation—Periodic movement back and forth between two extreme limits. An example is a plucked guitar string. A single vibration back and forth is one oscillation. A vibration also is described in terms of size (amplitude), oscillation rate (frequency), and timing (phase). In a suspension bridge, oscillation results from energy collected and stored by the bridge. If a bridge part stores energy beyond its capability, that part probably will fail.

Panel—The portion of a truss span between adjacent points of intersection of web and chord members.

Pier—Vertical support or substructure units supporting a multi-span superstructure at intermediate locations between the abutments.

Pile—A shaft-like linear member carrying loads through weak layers of soil or rock to deeper layers capable of supporting such loads.

Pile bent—A row of driven or placed piles, with a pile cap to hold in correct position; see Bent.

Plate girder—A large, solid web plate with flange plates attached to it by flange angles or fillet welds.

Portal—Open unobstructed space forming the entrance to a bridge.

Reinforced concrete—Concrete with steel reinforcing bars bonded within it to provide increased tensile strength and durability.

Resonance—Regular vibration of an object responding in step (at the same frequency) with an external force.

Rigid frame bridge—A bridge with moment (force-) resistant connections between the superstructure and substructure to produce an integral, elastic structure.

Riveted connection—A rigid connection of metal members assembled with rivets.

Safety hangers—Back-up for original connections to provide redundancy; often added for seismic retrofit.

Span—The distance between piers, towers, or abutments.

Stable—A structure's ability to resist forces that can cause material deformation or structural collapse.

Stay—Diagonal brace installed to minimize structural movement.

Steel—A hard, strong alloy of iron and carbon.

Stiffening truss—The sides of a suspended structure. A rigid, open frame that supports a road deck and consists of top chords, bottom chords, diagonals, and cross-frames.

Strand shoe—Device at the end of an eye-bar holding the wires of each strand for a main

suspension cable. The strands loop around the shoes to form a continuous cable from anchorage to anchorage.

Stringer—A longitudinal beam supporting a bridge deck.

Strut—A piece or member acting to resist compressive stress.

Substructure—The parts of a suspension bridge below water or earth; i.e., the piers, tower foundations, and the bottom portions of the anchorages.

Superstructure—The parts of a suspension bridge above water or earth; i.e., the towers, main cables, suspender cables, and related parts.

Suspended span—A simple span supported from the free ends of cantilevers.

Suspended structure—The portion of a suspension bridge suspended from main cables by suspender cables. Typically includes the road deck and stiffening truss.

Suspender cables—Vertical wires hung at regular intervals, connecting main suspension cables to the suspended structure. The connection is made with a cable band.

Suspension bridge—A bridge in which the floor system is supported by two main cables supported upon towers and anchored at their ends.

Tension—A pulling or stretching force.

Tie—A member carrying tension.

Torsion—Twisting force or action.

Tower—Vertical structure in a suspension bridge (or cable-stayed bridge) holding up the suspension cables.

Trestle—A structure consisting of spans supported upon frame bents.

Truss—A rigid, jointed structure made up of individual straight pieces arranged and connected, usually in a triangular pattern, so as to support longer spans.

Truss bridge—A bridge having a pair of trusses for the superstructure.

Upper chord—Top longitudinal member of a stiffening truss.

Viaduct—A series of spans carried on piers at short intervals.

Vortex—Rotary, swirling, or circular motion of wind or water. A vortex forms a vacuum at the center, drawing objects toward it. Examples are tornadoes and whirlpools.

Warren truss—Triangular truss with sloping members (and often vertical members) between top and bottom chords.

Web—The portion of a beam located between and connected to the flanges.

Web members—Intermediate members of a truss, not including the end posts, usually vertically or inclined.

Welded joint—A joint in which the assembled elements and members are attached by the fusion of metal.

BIBLIOGRAPHY

ARCHIVAL COLLECTIONS

California Institute of Technology Archives. Theodore von Karman papers.

Gig Harbor Peninsula Historical Society. Tacoma Narrows Bridge collection, and online exhibit "A Tale of Two Gerties."

National Archives and Records Administration, Washington, D.C. General records of the Federal Works Agency, project files, Tacoma Narrows Bridge, record group 162.2.3.

San Francisco Public Library, San Francisco History Center. Photograph collection.

Tacoma Public Library, Northwest Room. Photograph collection and newspaper articles.

University of Washington, Special Collections. Engineering Experiment Station records, F.B. Farquharson collection, and photographs.

Washington State Archives. Records of the Washington State Department of Transportation and the Washington State Toll Bridge Authority.

Washington State Department of Transportation, Library. Photographs, newspaper clippings, and other published information.

Washington State Historical Society. Historic photographs, newspaper clippings, and other published information.

UNPUBLISHED SOURCES AND REPORTS

Arvid Grant Associates. "Tacoma Narrows Bridge Condition, 1983" (conducted for the Washington State Department of Transportation).

_____. "Tacoma Narrows Bridge Report, August 1991" (conducted for the Washington State Department of Transportation).

Kawada, Tadaki. "Who Wrecked the Galloping Gertie? The Mystery of the Tacoma Narrows Bridge Disaster" (unpublished manuscript, 1975).

Locke, J. Allan. "The Development and Economic Significance of the Tacoma Narrows Bridge, 1923–1953. M.A. Thesis, College of Puget Sound, 1956.

Morelli, Dino A. "Some Contributions to the Theory of the Stiffened Suspension Bridge." Ph.D. Dissertation, California Institute of Technology, Pasadena, 1946.

Stevenson, T.A. "The Story of the Narrows Bridge and the People Who Made It Possible" (unpublished memoir), Northwest Room, Tacoma Public Library.

"Tacoma Narrows Bridge." Historic American Engineering Record, WA-99. Washington State Department of Transportation, Olympia, Washington, 1993.

"Tacoma Narrows Bridge, Tacoma Washington: Final Report on Design and Construction," by Clark Eldridge, 1940, Box 18, "Galloping Gertie" collection, WSDOT Records, Washington State Archives.

"Tacoma Narrows Bridge, Tacoma Washington: Report on Construction of the Substructure," by H.F. Connelly, 1940, Box 18, "Galloping Gertie" collection, WSDOT Records, Washington State Archives.

"Wheeling Suspension Bridge," Historic American Engineering Record, WV-2, Library of Congress.

INTERVIEWS

Brown, Margie
Clifford, Howard
Colby, Arnie
Durkee, Jackson
Elliott, Ed
Elliott, Darcie
Holcomb, Gerry Coatsworth
Howland, James
Laird, Linea
Linzy, Sharan
Matheny, Bill
Melson, Lewis B.
Moergan, Jon
Moore, Tim
Munson, Charles
Robeson, Jean
Storkman, Beverlee
Swinney, Richard L.
Taylor, Jeanette
White, Earl
Wylie, Kip

MOTION PICTURE FILM/VIDEO

The Camera Shop, Tacoma, Washington.

Farquharson, F. Burt, footage, University of Washington Libraries.

"Gertie Gallops Again." A 30-minute documentary, Tacoma Municipal Television, 1998 (copies available at several locations, including the Gig Harbor Peninsula Historical Society, Gig Harbor, Washington).

"Galloping Gertie." Barney Elliot interview by Enrique Cerna, *Evening Magazine,* KING-TV, Seattle, 1988.

Washington State Department of Transportation, footage in WSDOT records, Washington State Archives.

Published Sources

Articles and Essays

"Action of 'Karman Vortices'" (Blake D. Mills letter to editor). *Engineering News-Record* 125 (December 19, 1940): 808.

Addis, William. "Design Revolutions in the History of Tension Structures." *Structural Engineering Review* 6 (February 1994): 1–10.

"A Great Engineer" (editorial). *Engineering News-Record* 128 (September 9, 1943): 78.

Ammann, Othmar H. "Planning and Design of Bronx-Whitestone Bridge." *Civil Engineering* 9 (April 1939): 217–20.

_____, Charles A. Ellis, and F.H. Frankland. "Unusual Design Problems—Second Tacoma Narrows Bridge." *Proceedings of the American Society for Civil Engineers* 74 (June 1948): 985–92.

Andrew, Charles E. "Design of a Suspension Structure to Replace the Former Narrows Bridge—Part 1." *Pacific Builder and Engineer* 51 (October 1945): 43–45.

_____. "Redesign of Tacoma Narrows Bridge." *Engineering News-Record* 135 (November 29, 1945): 716–21.

_____. "Tacoma Narrows Bridge Number II…The Nation's First Suspension Bridge Designed to be Aerodynamically Stable." *Pacific Builder and Engineer* 56 (October 1950): 54–57, 101.

_____. "Unusual Design Problems—Second Tacoma Narrows Bridge." *Proceedings of the American Society of Civil Engineers* 73 (December 1947): 1483–97.

"Another Consultant Board Named for Tacoma Span." *Engineering News-Record* 125 (December 5, 1940): 735.

Averill, Walter A. "Collapse of the Tacoma Narrows Bridge." *Pacific Builder and Engineer* 46 (December 1940): 20–27.

Bashford, James. "Unusual Problems Attend Narrows Bridge Construction." *Compressed Air Magazine* (December 1939): 6033–35.

Berreby, David. "The Great Bridge Controversy." *Discover* (February 1992): 26–33.

Billah, Yusuf, and Robert Scanlan. "Resonance, Tacoma Narrows Bridge Failure, and Undergraduate Physics Textbooks." *American Association of Physics Teachers* 59 (February 1991): 118–24.

Billington, David P. "Creative Connections: Bridges as Art." *Civil Engineering* 60 (March 1990): 50–53.

_____. "History and Esthetics in Suspension Bridges." *Journal of the Structural Division, ASCE* (August 1977): 1655–72; (March 1979): 671–87.

"Board Named to Study Tacoma Bridge Collapse." *Engineering News-Record* 125 (November 28, 1940): 725.

Bowers, N.A. "Model Tests Showed Aerodynamic Instability of Tacoma Narrows Bridge." *Engineering News-Record* 125 (November 21, 1940): 674–77.

_____. "Tacoma Narrows Bridge Wrecked by Wind." *Engineering News-Record* 125 (November 14, 1940): 647, 656–58.

"Bridge Dedication Number: Lake Washington Floating Bridge-July 2, Tacoma Narrows Suspension Bridge-July 1." *Pacific Builder and Engineer* (July 6, 1940).

"Cable Spinning at Tacoma Narrows." *Engineering News-Record* 144 (February 16, 1950): 44–45.

Cissell, J.H. "Stiffness as a Factor in Long Span Suspension Bridge Design." *Roads and Streets* 84 (April 1941): 64, 67–68.

"Construction Features of the Tacoma Narrows Bridge." *Pacific Builder and Engineer* 52 (January 1946): 44–49.

"Dean of Europe's Radio War Broadcasters Is Brother of Lacey Murrow." *Pacific Builder and Engineer* 45 (November 4, 1939): 31.

"Details of Damage to Tacoma Narrows Bridge." *Engineering News-Record* 125 (November 14, 1940): 647.

"Details of Damage to Tacoma Narrows Bridge." *Engineering News-Record* 125 (November 21, 1940): 674–76.

"Details of Damage to Tacoma Narrows Bridge." *Engineering News-Record* 125 (November 28, 1940): 720.

"Dynamic Stability of Suspension Bridges" (editorial). *Pacific Builder and Engineer* 46 (December 1940): 1.

"Dynamic Wind Destruction" (editorial). *Engineering News-Record* 125 (November 21, 1940): 672–73.

Eldridge, Clark H. "An Autobiography: Capture and Imprisonment by Japs." *Pacific Builder and Engineer* 51 (December 1945): 44–49.

_____. "The Tacoma Narrows Bridge." *Civil Engineering* 10 (May 1940): 299–302.

_____. "The Tacoma Narrows Suspension Bridge." *Pacific Builder and Engineer* 46 (July 6, 1940): 35–40.

Embury, Aymar II. "Esthetics of Bridge Anchorages." *Civil Engineering* 8 (February 1938): 85–89.

"Fall of the First Tacoma Narrows Bridge." *Highway News* 12 (December 1964): 1–3.

Farquharson, F. Burt. "A Dynamic Model of Tacoma Narrows Bridge." *Civil Engineering* 10 (July 1940): 445–47.

_____. "Lessons in Bridge Design Taught by Aerodynamic Studies." *Civil Engineering* 16 (August 1946): 344–45.

"Film Showing Collapse of Tacoma Span Available." *Engineering News-Record* 125 (December 5, 1940): 733.

Finch, J.K. "Wind Failures of Suspension Bridges, or Evolution and Decay of the Stiffening Truss." *Engineering News-Record* 126 (March 13, 1941): 74–79.

"Galloping Gertie." *Newsweek* 16 (November 15, 1940): 23–24.

Gunns, Albert F. "The First Tacoma Narrows Bridge: A Brief History of Galloping Gertie." *Pacific Northwest Quarterly* 72 (October 1981): 162–69.

"High-Strength, Lightweight Deck for New Tacoma Narrows Bridge." *Engineering News-Record* 146 (January 11, 1951): 34.

Hills, Harold W. "The Techniques of Cable Spinning as Exemplified at the New Tacoma Narrows Bridge." *Western Construction* 57 (February 1951): 78–81.

Horwood, Ed. "Cable Spinning Operations Underway at Tacoma Narrows Bridge." *Pacific Builder and Engineer* 55 (November 1949): 44–47.

"'Jinx' Bridge Going Up Again." *Western Construction News* 24 (August 15, 1949): 61–63.

Jones, J. "Welded Cable Saddles for Tacoma Narrows Bridge." *Engineering News-Record* 123 (December 7, 1939): 91–92.

Johnston, Bruce, and H.J. Godfrey. "Test Model of the Tacoma Narrows Anchorage Bar." *Welding Journal* 18 (August 1939): 253–59.

"Lacey Murrow, Former Director, Honored at WSC." *Highway News* 8 (May–June 1959): 19.

Larsen, Allan. "Aerodynamics of the Tacoma Narrows Bridge—60 Years Later." *Structural Engineering International* 4 (2000): 243–48.

"Leon S. Moisseiff Dies: Famous Bridge Engineer." *Engineering News-Record* 128 (September 9, 1943): 70.

"Leon Solomon Moisseiff." In *Dictionary of American Biography*, Supplement Three, 1941–1945. New York: Charles Scribner's Sons.

MacPherson, A.R. "Construction Begins on New Tacoma Narrows Bridge." *Roads and Streets* 92 (January 1949): 63–65.

Mauldin, Douglas B. "'Galloping Gertie's' Legacy." *The Highway User* (April 1965): 21–23.

Melson, Lewis B. (letter to editor). *Oregon Stater* (Oregon State University) (February 1997).

"Memoir of Leon Solomon Moisseiff" (prepared by O.H. Ammann and Fredrick Lienhard). *Transactions of the American Society of Engineers* 111 (1946): 1509–12.

Miller, Bill. "Simmie." *Highway News* 2 (February 1953): 3-4.

Moisseiff, Leon. "Esthetics of Bridges" (review of book by Friedrich Hartmann). *Engineering News-Record* 105 (November 15, 1928): 741

———. "Growth in Suspension Bridge Knowledge." *Engineering News-Record* 123 (August 17, 1939): 46–49.

Murrow, Lacey V. "Construction Starts on the Narrows Bridge." *Pacific Builder and Engineer* 45 (March 4, 1939): 34–37.

"Narrows Nightmare." *Time* (November 18, 1940).

Neill, Thomas W. "Two New Washington Bridges that Write History." *Washington Motorist* 21 (July 1940): 4–5, 12.

"Pacific Northwest Bridges Completed." *Engineering News-Record* 125 (July 11, 1940): 2.

Peterson, Ivars. "Rock and Roll Bridge." *Science News* 137 (June 2, 1990): 344–45.

Petroski, Henry. "Leon Solomon Moisseiff." In *American National Biography*, Vol. 15. New York: Oxford University Press, 1999.

"Plywood Forms Are Used in Caissons for First Time on the Tacoma Narrows Bridge." *Pacific Builder and Engineer* 45 (May 6, 1939): 30–32.

"Puget Sound Bridge Proposed." *Engineering News-Record* 110 (February 2, 1933): 171.

Ross, Fred K. "Big Caisson Landed in Deep Tide Rip." *Pacific Builder and Engineer* 45 (June 3, 1939): 35–37, 48.

Ross, S., et al. "Tacoma Narrows, 1940." In *Construction Disasters*. New York: McGraw-Hill, 1984.

"Steel Tower at Tacoma Narrows Bridge Undamaged by $200,000 Blaze. " *Pacific Builder and Engineer* 55 (July 1949): 68.

"Tacoma Bridge Oscillations Being Studied by Model." *Engineering News-Record* 126 (April 24, 1941): 139.

"Tacoma Bridge Report Released by PWA Board of Consultants." *Engineering News-Record* 126 (March 27, 1941): 589.

"Tacoma Narrows Bridge Dismantling Recommended by Engineers." *Engineering News-Record* 126 (March 27, 1941): 453.

"Tacoma Narrows Bridge: Reconstruction to Follow Design Resulting from Extensive Wind Tunnel Research." *Roads and Streets* 90 (December 1947): 88–90.

"Tacoma Narrows Bridge Tolls Reduced." *Engineering News-Record* 125 (August 8, 1940): 197.

"Tributes to L.S. Moisseiff." *Engineering News-Record* 128 (September 23, 1943): 74–75.

"Why the Tacoma Narrows Bridge Failed." *Engineering News-Record* 126 (May 8, 1941): 75–79.

Books and Reports

Advisory Board on the Investigation of Suspension Bridges. *The Failure of the Tacoma Narrows Bridge*; a reprint of original reports. A contribution to the work of the Advisory Board on the Investigation of Suspension Bridges by the United States Public Roads Administration and the Agricultural and Mechanical College of Texas. College Station, Texas: School of Engineering, Texas Engineering Experiment Station, 1944.

America's Young Men: The Official Who's Who among the Young Men of the Nation. Los Angeles: Richard Blank, 1934.

Andrew, Charles. *Final Report on Tacoma Narrows Bridge.* Tacoma: Washington Toll Bridge Authority, 1952.

Cheney, Sheldon. *Art and the Machine.* New York: McGraw-Hill, 1936.

Condit, Carl W. *American Building Art: The Twentieth Century.* New York: Oxford University Press, 1961.

Farquharson, F. Burt. *Aerodynamic Stability of Suspension Bridges with Special Reference to the Tacoma Narrows Bridge.* Seattle: University of Washington Press, 1954.

Feld, Jacob, and Kenneth Carper. *Construction Failure*, 2nd Ed. New York: John Wiley and Sons, 1997.

Gies, Joseph. *Bridges and Men.* Garden City, New York: Doubleday, 1963.

Gotchy, Joe. *Bridging the Narrows.* Gig Harbor, Washington: Gig Harbor Peninsula Historical Society, 1990.

Hadlow, Robert W. *Elegant Arches, Soaring Spans: C.B. McCullough, Master Bridge Builder*. Corvallis: Oregon State University Press, 2001.

Hopkins, Henry James. *A Span of Bridges: An Illustrated History*. New York: Praeger, 1970.

Levy, Matthys, and Mario Salvadori. *Why Buildings Fall Down: How Structures Fail*. New York: W.W. Norton, 1992.

McCullough, Conde B., Glenn S. Paxson, and Dexter R. Smith. *An Economic Analysis of Short-span Suspension Bridges for Modern Highway Loadings*. Technical Bulletin No. 11. Salem: Oregon State Highway Department, 1938.

————. *The Derivation of Design Constraints for Suspension Bridge Analysis (Fourier-series Method)*. Technical Bulletin No. 14. Salem: Oregon State Highway Department, 1940.

————. *Rational Design Methods for Short-span Suspension Bridges for Modern Highway Loadings*. Technical Bulletin No. 13. Salem: Oregon State Highway Department, 1940.

Menzies, Archibald. *Menzies' Journal of Vancouver's Voyage, April to October 1792*, Archives of British Columbia, Memoir No. V, Edited by C.F. Newcombe. Victoria, B.C.: William H. Cullin, 1923.

Morgan, Murray, and Rosa Morgan. *South on the Sound: An Illustrated History of Tacoma and Pierce County*. Woodland Hills, California: Windsor, 1984.

"Official Opening: Tacoma Narrows Bridge and McChord Field, June 30–July 4, 1940, A.D." (dedication program). Tacoma: Johnson-Cox Company [WSDOT records, "Galloping Gertie Collection," Washington State Archives].

Petroski, Henry. *Design Paradigms: Case Histories of Error and Judgement in Engineering*. New York: Cambridge University Press, 1994.

————. *Engineers of Dreams: Great Bridge Builders and the Spanning of America*. New York: Alfred A. Knopf, 1995.

————. *To Engineer Is Human: The Role of Failure in Successful Design*. New York: Vintage, 1992.

Plowden, David. *Bridges: The Spans of North America*. New York: W.W. Norton, 1974.

Ratigan, W. *Highways over Broad Waters: Life and Times of David B. Steinman, Bridgebuilder*. Grand Rapids, Michigan: William B. Eerdmans, 1959.

Robinson, John V. *Al Zampa and the Bay Area Bridges (Images of America)*. San Francisco: Arcadia, 2005.

Russell, Steven J. *Kalakala: Magnificent Vision Recaptured*. Seattle: Puget Sound Press, 2002.

Scott, Richard. *In the Wake of Tacoma: Suspension Bridges and the Quest for Aerodynamic Stability*. Reston, Virginia: ASCE Press, 2001.

Smith, Martin J., and Patrick J. Kiger. *OOPS: Twenty Life Lessons from the Fiascos that Shaped America*. New York, Harper Collins, 2006.

Stackpole, Peter. *The Bridge Builders: Photographs and Documents of the Raising of the San Francisco Bay Bridge, 1934–1936*. Corte Madera, California: Pomegranate Artbooks, 1984.

Steinman, David B. *Songs of a Bridge Builder*. Grand Rapids, Michigan: William B. Eerdmans, 1959.

————, and Sara Ruth Watson. *Bridges and Their Builders*. New York: G.P. Putnam's Sons, 1941.

"Supplemental Tests on the Dynamic Model of the Original Tacoma Narrows Bridge." University of Washington, Structural Research Laboratory, 1943.

Talese, Gay. *The Bridge*. New York: Walker, 1964.

"Toll Bridges and Ferries: Tacoma Narrows Bridge." "Washington State Highway Commission First Biennial Report"; "Washington Department of Highways Twenty-fourth Biennial Report, 1950, 1952." Olympia: State Printing Plant [1952].

Van der Zee, John. *The Gate: The True Story of the Design and Construction of the Golden Gate Bridge*. New York: Simon and Schuster, 1986.

Van Leeuwen, Thomas A.P. *The Skyward Trend of Thought: The Metaphysics of the American Skyscraper*. Cambridge, Massachusetts: MIT Press, 1988.

Who's Who in the State of Washington, 1939–1940. Seattle, 1940.

Wilkes, Charles. *Narrative of the United States Exploring Expedition...*, Vol. IV. Philadelphia, 1850.

Newspapers

Daily Olympian
Seattle Post-Intelligencer
Seattle Times
Tacoma Daily Ledger
Tacoma News Tribune
Tacoma Sun
Tacoma Times

Selected Web Sites

Bristol University (United Kingdom).
www.enm.bris.ac.uk/research/nonlinear/tacoma/
tacoma.html

Design Failure Lessons, University of Texas, Department
of Mechanical Engineering.
www.me.utexas.edu/~me179/topics/lessons/
case1.html

Gig Harbor Peninsula Historical Society and Museum
("A Tale of Two Gerties: How and Why We Bridged the
Narrows," online exhibit).
www.gigharbormuseum.org/nbonlinexhibit.html

Tacoma Narrows and suspension bridges on PBS, *Nova,*
online.
www.pbs.org/wgbh/nova/bridge/meetsusp.html

Tacoma Public Library (photograph database of approximately 200 images for the 1940 and 1950 bridges).
search.tpl.lib.wa.us/images/

University of Washington Libraries (online exhibit of
190 images presenting the story of the construction, collapse, and rebuilding).
www.lib.washington.edu/specialcoll/exhibits/tnb/

Video film clip of the collapse.
www.camerashoptacoma.com

WSDOT—Narrows Bridge Project (updates for the
2007 bridge).
www.wsdot.wa.gov/projects/sr16narrowsbridge/

WSDOT—Tacoma Narrows Bridge History.
www.wsdot.wa.gov/TNBhistory

Index